W9-BLW-484

OUCH!

Also available in the Bloomsbury Sigma series:

Sex on Earth by Jules Howard
Spirals in Time by Helen Scales
A is for Arsenic by Kathryn Harkup
Herding Hemingway's Cats by Kat Arney
Death on Earth by Jules Howard
The Tyrannosaur Chronicles by David Hone
Soccermatics by David Sumpter
Big Data by Timandra Harkness
Goldilocks and the Water Bears by Louisa Preston
Science and the City by Laurie Winkless
Bring Back the King by Helen Pilcher
Built on Bones by Brenna Hassett
The Planet Factory by Elizabeth Tasker
Wonders Beyond Numbers by Johnny Ball
Making the Monster by Kathryn Harkup
Catching Stardust by Natalie Starkey
Seeds of Science by Mark Lynas
Eye of the Shoal by Helen Scales
Nodding Off by Alice Gregory
The Science of Sin by Jack Lewis
The Edge of Memory by Patrick Nunn
Turned On by Kate Devlin
Borrowed Time by Sue Armstrong
We Need to Talk About Love by Laura Mucha
The Vinyl Frontier by Jonathan Scott
Clearing the Air by Tim Smedley
Superheavy by Kit Chapman
18 Miles by Christopher Dewdney
Genuine Fakes by Lydia Pyne
Grilled by Leah Garcés
The Contact Paradox by Keith Cooper
Life Changing by Helen Pilcher
Friendship by Lydia Denworth
Death by Shakespeare by Kathryn Harkup
Sway by Pragya Agarwal
Bad News by Rob Brotherton
Unfit for Purpose by Adam Hart
Mirror Thinking by Fiona Murden
Kindred by Rebecca Wragg Sykes
First Light by Emma Chapman

OUCH!

WHY PAIN HURTS AND WHY IT DOESN'T HAVE TO

Margee Kerr and
Linda Rodriguez McRobbie

BLOOMSBURY SIGMA
LONDON • OXFORD • NEW YORK • NEW DELHI • SYDNEY

BLOOMSBURY SIGMA
Bloomsbury Publishing Plc
50 Bedford Square, London, WC1B 3DP, UK

BLOOMSBURY, BLOOMSBURY SIGMA and the Bloomsbury Sigma logo
are trademarks of Bloomsbury Publishing Plc

First published in the United Kingdom in 2021

A catalogue record for this book is available from the British Library

Library of Congress Cataloguing-in-Publication data has been applied for

ISBN: HB: 978-1-4729-6527-1; TPB: 978-1-4729-6529-5;
eBook: 978-1-4729-6525-7

2 4 6 8 10 9 7 5 3 1

Typeset by Deanta Global Publishing Services, Chennai, India
Printed and bound in Great Britain by CPI Group (UK) Ltd, Croydon CR0 4YY

Bloomsbury Sigma, Book Fifty-nine

Contents

INTRODUCTION

Pain (Probably) Isn't What You Think

We've all got a story about pain. Maybe it's that time you broke your arm skating, skiing or just walking too fast. Or the time you finished the game on a busted ankle or the 10 hours of labour without an epidural. Maybe your story of pain is a story of violence, the injury and trauma of an assault. Maybe it's a story of terror, of the accident and what happened after. Or it's heartbreak, the seemingly endless depths of grief and despair after a loss. Whatever it is, (almost) all of us have experienced what we call pain and we're not in a hurry to experience it again.

But have you ever tried to define that pain? When you're telling this story, how do you explain the pain? Do you try to quantify the injury – how many broken bones, the size of the bruise, the amount of blood? Or do you describe the cause – the type of cancerous cells, the crowning baby, the sharp knife? But what if there was no obvious cause? And how do you communicate the intensity? Is it a searing or scalding burn, a throbbing or dull pressure, a pounding or stabbing headache? Is it worse than a bee sting but not as bad as a dog bite? Is your pain a 7 heading for an 8 on a scale of 10?

There's a reason why talking about pain turns into a story; for a supposedly universal experience, how pain is felt can be surprisingly difficult to convey. And that's because the ideas we each of us hold about pain, though similar, aren't all the same. Many of our stories about pain are informed by other stories about pain, by cultural, social, medical, political narratives that define pain in ways that may or may not actually be all that useful. For starters, the typical narrative about pain says that we avoid it. But think for a moment and you can come up with dozens of examples of when we actively *don't* avoid pain. Consider that 20 per cent of all Britons have paid someone to repeatedly jab an ink-filled needle into their skin. That percentage rises precipitously among younger generations – nearly half of all millennials in the UK have a tattoo. Upwards of 80 per cent of

American women have their ears pierced. Millions of us pay people to inject or stuff substances under our skin to plump up areas we think are deficient or to break our noses into a finer shape. Millions more endure a bit of sudden agony to preserve the fiction that we are hairless in the bits that aren't 'supposed' to have hair.

Have you ever bitten your lip or pinched your arm to keep yourself from saying something mean? Or given yourself a hard slap to keep from dozing off at the wheel? And what about the kind of pain we think is required to 'gain'?* The incidence of running-related injuries is difficult to nail down – studies have found that anywhere from 20 to 92 per cent of runners suffer injuries related to their sport, and that's all down to the definitions of 'runner' and 'injury'. Suffice to say, however, when we tie on our running shoes, most of us know that every step is going to cause some kind of pain somewhere, and we do it anyway. But sport is hardly the only time we weigh the balance between pain and pay-off and decide it's worth it – a significant number of women choose to go through what is supposed to be one of the most painful experiences more than once and it's not because they 'forgot' the pain of childbirth.

For some, pain might just be the price to pay for something else they want, sure. But for many, pain *is* the point. Maybe your pain story is a bit more, well, pleasant. Humans can enjoy pain – and we're not just talking about the hundreds of thousands of people who fancy a little BDSM in the bedroom. More than 2 billion people in the world love – *love* – the eye-watering burn that comes from eating hot chilli peppers and spicy curries. Every New Year's Day, hundreds of thousands of people around the world strip to their swimsuits and plunge into their nearest icy waters for a painful, exhilarating rush (and believe us, cold water *hurts*). And the last decade has seen a tremendous rise in the popularity of obstacle-course races that explicitly promise you'll emerge from the mud battered and maybe – hopefully – even bloody.

What about the idea that pain is a corrective? Pain *does* teach us to not touch the hot pan again, that's true – but we can all remember times

---

* Thank you, Jane Fonda. Though the sentiment predates her by at least 225 years (Benjamin Franklin noted in 1758 that 'There are no gains without pains'), Fonda's 1980s workout videos introduced 'no pain, no gain', along with 'feel the burn', to a wider audience.

that we ignored what we learned. It took pro-skateboarder Tony Hawk 10 years and multiple concussions, broken ribs and lost teeth to become the first person ever to land the 900, a two-and-a-half-revolution aerial spin, at the 1999 X Games. And while there's some glory in suffering for your sport, consider the hangovers you might have had in your lifetime. Notice we said 'hangovers'. As in more than one.

We also tend to assume pain is synonymous with injury, and for good reason – pain is a very useful warning system. But not always. Some things that are actually damaging to the body don't hurt at all; we don't realise we're sunburned until the skin cells begin to die. Sometimes this warning system is working with, well, compromised information – consider the story of the unfortunate builder who stepped down on a 15cm nail. The nail punctured his steel-toed boot; even the smallest movement caused unbearable pain. After the builder was dosed up with fentanyl (a synthetic opioid stronger than heroin) and midazolam (a benzodiazepine-based psychoactive used to calm agitation), the nail was pulled from his foot. 'When his boot was removed a miraculous cure appeared to have taken place,' the 1995 report in the *British Medical Journal* reads. 'Despite entering proximal to the steel toecap, the nail had penetrated between the toes: the foot was entirely uninjured.' (There's no mention of the builder's pride, however.)

The nail puncture that wasn't demonstrates the effect expectation has on our perception of pain – we can create pain even in the absence of painful stimuli. This is the 'nocebo' effect, the evil twin of the more well known placebo effect. Researchers can induce this effect by employing the 'rubber hand illusion': people come to feel ownership of a mannequin hand after a series of visual and tactile stimulations. When the rubber hand is burned, subjects report painful sensations in their real hand. And they feel better when a topical analgesic was applied, again, *to the fake hand*. That all might seem like some kind of perceptual sleight of hand (sorry), but there's more evidence that pain doesn't need a body: some estimates put the incidence of phantom-limb pain, experiencing pain in limbs that are no longer attached to the body, as high as 90 per cent. According to those same estimates, 5 to 10 per cent of sufferers described their pain as severe.

So pain isn't always a reliable indicator of harm, nor does it always have an obvious cause – one in four people who visit their GP in the

UK describing a pain in their body somewhere will come away with a diagnosis of 'medically unexplained symptom' that test after test simply can't explain. But truthfully, a lot of pain is confusing and medical science, though monumental, doesn't have all the answers. Most tissue damage heals within three months, but as much as 20 per cent of the global population is living with pain that lasts far longer than that. Chronic pain, say some researchers, can be a bit like an alarm system that's stuck, relaying the same message over and over, to the point that it's no longer providing any useful information.

Nor is our physical body – real or phantom – the only source of pain. Though Takotsubo cardiomyopathy (also known as broken heart syndrome) is fairly unusual, it's common enough to show up on the NHS's website for potential heart conditions. Acute stress – emotional disturbance – can cause the heart's left ventricle to actually change shape, producing severe chest pain and breathlessness. Many of us know from experience that grief, remorse and rejection hurt. This pain can be felt more intensely than a broken arm and can disrupt our lives just as much, if not more – in fact, the neural patterns of physical pain sensations and social rejection are strikingly similar.

Pain is anything but straightforward and trying to communicate it presents its own issues. When we say 'pain', we tend not to be specific, other than to say where it hurts (and even 'where' can be hard to pinpoint). We say a root canal is painful, as is a cut finger, as is chemotherapy, as is arthritis, as is muscle ache, as is eating hot peppers, as is a broken heart, yet these experiences are wildly different from one another. Their single link is that we use this one astonishingly flexible but utterly insufficient word to describe them all. A word that, when we try to define it succinctly, sounds a lot like US Supreme Court Justice Potter Stewart attempting to define obscenity in 1964: 'I know it when I see it.'

The more we question this experience that we call pain, the more we are confronted by the fact that pain is complex, hard to describe and even harder to understand. Maybe we don't actually know pain when we see it.

What pain means matters

Pain is a big concept, no matter how we flatten the language we use to talk about it. It is constructed not only from anatomical structures and neurochemical phenomena, but also where we are, who we're with, the reason we're there, our previous experiences with pain, what we expect to feel, and what we want. How we react in this moment and how we think about it in retrospect determine whether it will become part of the rush or the trauma. And each time we experience pain, that experience shapes our response to the next time we'll experience it, and that time shapes the next, and so on. It is shaped by our genetic material, the physical environment in which we're born and raised, and our personal morals and values, which in turn are forged in and framed by the culture, religion and politics of our time.

And right now, we are suffering from the symptoms of a socially dysfunctional relationship with pain. Pain is a multi-layered and complex phenomenon, yet the way we treat it is not: avoidance and suppression, typically through drugs. We have more ways to pharmacologically manage pain, but opioids and over-the-counter analgesics often cause more problems than they solve. We're not only talking about the opioid epidemic – the increasing availability of ever more powerful drugs means that more and more, we expect to be pain-free. And when we aren't, this has some very serious consequences for our health and happiness. The irony is that the more we try to suppress pain, the more we feel it.

It's not just the drugs that promise to deliver this pain-free existence. Powerful forces – from big pharmaceutical companies to Instagram to the relentless narrative of consumerism – tell us that we can feel good, that we deserve to feel good, and that we should feel good all the time. Ideologically, we are increasingly circling the wagons – self-curated social media and news feeds mean we don't have to be hurt by ideas we don't like – angered, sure, but not hurt – And we're making sure the same will be true for the next generation. Some of us might be labelled helicopter or snowplough parents, but when it comes to children, the category of 'harm' has dramatically expanded in the last 40 years to include not only real threats but also the possibility of discomfort. There is no place for pain in today's comfortable paradigm.

We're not the only ones saying this. 'The 21st-century world we live in can be characterised as an "analgesic culture", one in which we work to avoid pain and distress,' Christopher Eccleston, Director of the Centre for Pain Research at the University of Bath, wrote in an article for the British Psychological Society blog. 'When the avoidance of pain fails our first thoughts are that any pain should be short-lived, diagnostically relevant, treatable, and a cause for empathy, sympathy or social assistance.' When we experience pain that doesn't meet that criteria, that pain hurts more; the flaw in our relationship with pain is based on our expectation that we shouldn't have to suffer it.

In the 1980s, Harvard psychiatrist Arthur Barsky warned that America was becoming a nation more sensitive to pain, and offered some convincing data to back up his claim. He noted that where community surveys from the 1920s found respondents had 0.82 episodes of serious illness a year, by the 1980s this increased to 2.12 episodes, and that these episodes lasted longer. Even after accounting for increases in awareness and life expectancy, the differences were significant. Americans were objectively healthier, yet they said they felt worse. His argument – and he was not the first or last to make it – was essentially that our tolerance for discomfort decreased as our expectation to be comfortable, to be pain-free, increased.

In our defence, this wasn't an entirely unreasonable expectation – after all, during the 20th century, we developed treatments and vaccines for many acute and infectious illnesses, came up with new and exciting pharmacological ways to address pain, our life expectancy doubled, and the safety of our homes and workplaces increased. But as hard as pain is to define, it's equally difficult to snuff out completely – not even the strongest opioids can reliably do it. This mismatch between expectation and reality has darkened our perception of the pain we're in and has made it feel worse.

More than 30 years later, the trends Barsky observed appear to have grown. In 2017, the US National Bureau of Economic Research, in a paper analysing survey data from 2011, found that Americans reported experiencing aches and pains more often than any other nation, developed or developing. According to the survey, 34.1 per cent of Americans reported feeling physical pain 'often' or 'very often';

Australia, at 31.7 per cent, was closely followed by the UK, at 29.4 per cent. At the same time, the US spends more money on healthcare than any other nation, at about $11,172 per person in 2018. But again, Americans say they feel worse. Speaking to *The Atlantic* about the data, Barsky suggested that Americans assume all aches and pains can and should be treatable, and that it would therefore be intolerable to suffer them. 'Curable pain is unbearable pain,' he told the magazine. 'It's when you think you shouldn't have to suffer it, that there should be some solution out there, that it becomes even more intolerable.'

As we adjust our lives around avoidance and suppression, we internalise the message that we cannot handle pain. And when we limit our chances to get hurt, we fail to learn that we can get back up again. This has serious, demonstrable consequences for our ability to deal with both the physical and emotional pain that life will inevitably throw at us, and fuels a paradigm in which we don't believe we have control over pain without the aid of drugs, surgery or medical intervention.

Our reliance on drugs and surgery is an unintended consequence of the incredible advances in medicine that have ended some of the worst forms of human suffering. The dominance of the biomedical model of the human body for the better part of the last four centuries, helped foster those advances but it has left us with a big blind spot when it comes to understanding and managing pain. This model considers the human body as made up of constituent parts that can be assessed and repaired – just find the broken bit and fix it. Treating the human body like a car can be useful, sure, but this model of human experience ignores pain that doesn't fit into the 'broken part' model and it ignores the role of emotion and cognitive processing in the generation and management of pain. It also means that too many of us still think some kinds of pain are more 'real' than others, and it's the reason that when someone says 'It's all in your head', it's not typically meant kindly.

The invented divide between 'emotional' and 'physical' pain is perhaps the biggest misconception about pain that we need to unlearn. Our emotional states have a demonstrable impact on our physical state and vice versa because they are not distinct entities, not even two sides of the same coin but the actual coin itself. The dominant and overly simplistic approach to pain also has ramifications for how we relate to the painful experiences of other people. If we barely understand our own pain, how can we be expected to perceive another person's phenomenological state? We are all – even those in the medical profession – guilty of underestimating or even dismissing other people's pain because it doesn't match our own or what the textbooks say or what we think it *should* look like.

The artificial division of mind and body has not only had serious consequences for how we diagnose and see pain, it has also meant that potential pathways for easing pain have been ignored. For example, a 2013 study published in the journal *Pain* found that when the meaning of a painful experience was reframed from detrimental to beneficial, participants exhibited a much higher pain tolerance. But what was more interesting was the fact that this increased tolerance seemed to have been aided by the co-activation of the opioid and cannabinoid systems, our endogenous painkillers; how we think about our painful experiences, what they mean to us, has measurable neurobiological effects that change how we feel pain. In one of our favourite examples, swearing or saying 'bad words' can lessen the perception of pain (sadly, the effect is reduced the more you swear – habitual swearers don't get nearly the same analgesic relief as people who don't swear at all).

Of course, as relieving as a well-timed '*F*ck!*' can be in the short term, we can't swear our way to a better relationship with pain. That's going to take a lot more work.

Our stories about pain

Like we said at the beginning, everyone has a story about pain. And we wanted to write this book because we think there's a lot we can learn through listening to and telling stories about pain.

We have both experienced pain; we're human. Some of those experiences overlap – when we first talked about the idea for the book, we realised we'd both pierced our own navels as teenagers. It was the 1990s, what can we say? Margee used an ice cube and a safety pin; Linda used an ice cube and a needle. Both experiences were sort of painful, but not really, and neither amateur piercing lasted. But in telling these stories of essentially the same act, we recognised that we each had a different relationship to pain, that Margee's phenomenological perception of piercing her belly button was essentially different to Linda's.

Our personal relationships to pain have, of course, changed since we were needle-happy teens. The stories we tell ourselves about our painful experiences – the nerve pains and the monthly migraines, the crippling anxiety and depression and the chronic gastrointestinal distress – and about what they mean has shaped how we feel about the pain we are inevitably in. When we are in pain, we are typically at our most vulnerable, most naked; it's also the moment we can learn new truths about ourselves and each other. These stories are what motivated us to understand how our experiences of pain are created and constructed, what mitigates or exacerbates them. We wanted to know whether we could learn to weather these experiences better – and whether there might even be an upside to them. (Spoiler: we did and there is!)

Before we move on, then, we want to tell you some of our pain stories. These are the stories that led us here, that have shaped who we are and how we think about pain. The pain we lived through will not be the kind of pain that you experience, of course, and our perception of it is coloured by our memories, our knowledge, our own pain thresholds and tolerance for discomfort, not to mention genetics, epigenetics, childhood and adulthood traumas, and the fact that we are writing a book about how to make friends with pain and have drunk our own Kool-Aid. But we know that pain can be scary. If we're going to be your guides in figuring out why it shouldn't always be, we want you to know who we are.

Margee's pain story

I knew I wanted to write a book on pain in early 2015, right after I had finished my book *Scream: Chilling Adventures in the Science of Fear.* Throughout my research, I quickly learned that inflicting pain, typically through non-harmful, but certainly painful electric shocks, was the go-to approach in studying fear in the lab. For example, if you want to measure a person's reaction time under stress, you first have to give them something to stress about. As anyone who has trained their dog with a shock collar can tell you, electricity – even just the promise of an electric shock – can be scary. But I also knew that pain and fear are not the same, and that not all pain was scary. As with fear, my own experiences reminded me that pain isn't always the *big bad* and I kept coming back to the question of how these two dynamic processes were related.

Growing up riding horses and getting lost in the woods outside Baltimore offered up many opportunities for pain, from broken fingers and toes courtesy of various hooved animals, to nasty bruises, cuts and skinned knees, and even a big ol' swollen black eye after a particularly nasty fall from my horse, Cookie. In my late twenties, I played roller derby – a sport practically defined by pain, either trying to inflict it, avoid it or survive it. The pain after my first practice left me hobbling around my research office for a good week. But the pain felt good, like sore muscles after a challenging run, only with more bruises. Even the injuries made me feel strong, each a testament to our ability to endure and to heal. And it wasn't the couple of concussions or broken wrist that led me to retire; I just woke up one day and didn't want to hit people any more. (I'm still a big fan of the sport, though.)

Pain in pursuit of adventure was a fair price for me; 'no pain, no gain' made sense to me. Protestant ethics of industry and responsibility were well internalised, and I welcomed any kind of evidence that I was challenging myself, that I was working. This is all to say that I thought I had a high pain threshold and a high tolerance, and I took pride in that.

I also knew the unwanted and unwelcome kinds of pain, the worst of it coming from my own mind. I know the names for it now – negative intrusive thoughts and rumination – but for a long time I thought everyone's mind worked like mine and I was just worse at managing it. I still remember, very clearly, a night of tossing and turning because I couldn't stop thinking about a wrong answer on a maths quiz in the third grade. As an adult, negative experiences would become well-trodden memories, painfully replayed every night, turned over and over as I tried to figure out how they could have gone better. With more years in therapy than out, and as a connoisseur of antidepressants starting at age 15, I accepted long ago that with anxiety and depression as my sidekicks, pain would never be too far behind.

By the time I sat down to get to work on this book, it was very much from the perspective of 'pain – *pshaw* – I got this.'

Then one morning, Friday 28 August 2015 to be exact, I woke up in intense pain. The muscles in my neck and shoulders felt like they had been superglued. Even the slightest movement exploded an excruciating sensation of ripping and burning at the same time. And then there was my left arm, laying at my side, dead. At least, that's what it looked like; I couldn't feel it at all. I stared at it for about a minute: heavy, dull, dead. Then, like a zombie coming to life, bursts of electricity started twitching through my arm, building into a fire and then settling into a clenched achiness, as though it were compressed in a vice.

Pain. In no way had I 'got this'.

For the first time in a long time, I was scared. I headed straight for my doctor, who suspected a muscle spasm, which I'd experienced before. But I *knew* it was different and begged for an x-ray. The 'dead arm' pushed her over the edge and she agreed, ordered the x-ray and gave me a prescription for the muscle relaxer baclofen and 500mg of naproxen. I went right from her office to radiology, and within an hour she called back. I could hear the surprise in her voice as she told me there was evidence of bone degeneration in my cervical spine. She was ordering an MRI and suggested I make an appointment with a surgeon.

In the following month, I would learn I had osteoarthritis of the spine and cervical stenosis with radiculopathy, a condition that mostly shows up in those over the age of 65. (*Radiculopathy*, I thought, sounded like a Mike Judge movie, and made me smile.)

During a brief 10-minute consultation, my neurosurgeon explained the condition: a cervical disk was essentially being squashed between two deteriorating vertebrae, crowding the nerve pathway between the C5 and C6. These were the facts. He said to call him when I was ready for surgery, which could be tomorrow or in five years. In the meantime, I could try a steroid epidural, which might reduce the inflammation and give my claustrophobic nerve some space (it didn't).

My doctor's brevity didn't bother me; I appreciated the mechanical approach to what I saw as a mechanical problem. Looking back, I should probably have been sad, but all I remember thinking was 'yep, that sounds about right.' Arthritis is pervasive among my family, and I hadn't necessarily been easy on my body, to say the least. But at the time I didn't really think – or maybe I didn't let myself think – about how this mechanical problem would impact my more than mechanical self.

Out of stubbornness and denial, it would take me five months to make the call to schedule a spinal fusion surgery. During that time I was continually surprised by my sensations; nothing I had experienced compared to this new infuriating, devastating nerve pain. I say pain now, but it wasn't *really* pain, at least not the kind I was familiar with. To me it was a never-ending flow of intense tingling down my arm and into my fingers and a relentless aching. It was twitching muscle spasms and sleepless nights. To avoid the absolute worst of it, I had to keep my head tilted back and to the right at about a 45-degree angle, and my left arm up over my head and slung behind my back in the most awkward yoga pose ever.

My worst pain memory to this day is making it through the 45-minute MRI. I'm not one to 'tap out' of anything and I'm not claustrophobic, but remaining inside the MRI machine – laying perfectly still, flat on my back, arms to my sides, unable to contort into my awkward yoga pose – is one of the most difficult pains I've endured. Immediately, burning shockwaves shot down my arm as though gasoline had replaced my blood. I imagined my nerve tract, pinched between two crumbling vertebrae, screaming for space, and there was nothing I could do about it. No amount of reframing or renaming would work; I tried visualising myself on the hills with the Von Trapp family, running across the mountains, singing about my favourite things, trying to solve the problem of Maria. I tried to forget my body. At some point I guess I started holding my breath, and the attendant reminded me through the intercom to breathe.

My pain-management doctor asked if I wanted a prescription for oxycodone, but I quickly declined. For one thing, I am terrified of addiction and for another, I knew, thanks in part to writing this book, that opioids were not the solution for my kind of pain. Where I did find relief was in treating pain with pain using transcutaneous electrical nerve stimulation, or a TENS unit. Growing up on farms, my exposure to the power of electricity was early and intense: 'Margee, don't touch the electric fence, it will shock you.' Me: touches electric fence. After that, I no longer wondered how a thin wire could contain a whole herd of cows. For months I walked around wired up like Frankenstein's monster. It was incredibly effective, but not because it *stopped* the incessant tingling and twitches, it just disrupted and replaced them with signals of my choosing, delivered in predictable patterns at a slightly higher intensity.

It took a lot to finally motivate me to make the call to schedule surgery. The intensity of pain can force us to confront realities we're not ready, don't know how or are simply unwilling to face. For me, that was building a relationship with my body, one that was not quite so antagonistic. I know now that, no matter how hard we might try to avoid and deny the unpleasant sensations (and I tried pretty hard), in the end the only way out is through.

In the six months between waking up with a zombie arm and waking up with a cadaver bone graft, my understanding of pain and of myself expanded and deepened profoundly. I began this writing with a laser focus on the upside of pain, a focus that admittedly became difficult to keep as I experienced a deluge of horrific sensations I didn't believe possible for the human body to experience. And my pain story did not end with my surgery; adaptive habits formed in times of pain are fast learned and hard to break, even when they become the problem rather than the solution. But that is a pain story for another chapter.

Through all of this I learned that being able to disconnect from our bodies and to suppress pain may be adaptive in the short term but simply doesn't work in the long run, and it is not synonymous with strength or outdated notions of bravery. Letting the pain in, feeling it all, can be terrifying but that is when healing and growth can happen. Choosing to take that step, now that is an act of bravery.

Linda's pain story

When Margee first reached out to me with the idea for this book, I started thinking about my own experiences with pain. Of course

I did – this is what happens when we start talking about pain. I thought about the first time I used the NHS, after I'd got my finger stuck in the beater of a hand mixer and accidentally turned it on. As an American, I was surprised and confused to leave A&E with a grotesquely swollen finger in a splint but without either a bill to pay *or* a prescription for exciting pain-relieving drugs. I was even more surprised to find out that really, I didn't need the drugs.

That was the moment that made me start wondering whether the drugs were really all that necessary, at least in some circumstances, that maybe the narrative I'd been raised on was wrong. But when I think about painful experiences of the kind that give shape and dimension to who you are, the first thing I think about is that I carried and delivered two children. And this utterly quotidian miracle went a long way towards making me recognise that I am bigger than the pain.

The experiences of delivering these two boys – both about the same size, both making their entry into the world via the same vagina – could not have been more different. With Austin, my first, I spent the first 17 hours of labour hanging out. I baked a batch of corn muffins – gripping the edge of the counter through each contraction – watched trash TV, drank a half-pint of Guinness, took two hot showers, popped some paracetamol. This was all right, I told myself. And for a while, it was. But shortly before we left for the hospital, I hit the floor. It was suddenly like I was being cut in half by a flaming buzz-saw. As waves of pure fire scorched my lower back, I panted on my hands and knees, tears squeezing from my eyes. This was *not* all right.

When we arrived at the hospital, I was faced with a 'choose your own adventure' decision: go to the fifth floor for the labour and delivery ward, where I could get an epidural, or go to the eighth-floor birthing centre, where it would be all yoga balls and hot tubs but no epidural. At that moment, I was scared. The idea of more of *this* made me feel like I would come out of my skin with real, visceral panic. I could not see any way of getting past this pain. I went for the epidural. By the time the anaesthetist arrived to give it, I was shivering violently – a common occurrence in labour that I'd had no idea was going to happen – and afraid I wouldn't be able to hold still long enough for him to slide the needle into my spine. I did and he did. The pain subsided until it was a ghost of itself. Ten hours later, I held my sweet son to my chest while the midwife stitched up what had torn down at the business end. I couldn't feel it.

My experience wasn't traumatic but it could have gone better. The epidural slowed down labour, and when the time came to push, I had so little feeling that I wasn't aware of the damage I was doing to my pelvic floor. I'd pushed so hard that my face and chest were freckled with broken blood vessels, as if I'd stood on my head for too long. I'd pushed so hard that stuff that was supposed to stay on the inside didn't. And then there was the pain thing. I didn't blame myself for getting the epidural – I didn't need to be a hero, I thought, I just needed a baby – but I was left wondering 'what if'. What if I hadn't had the epidural? Would the pain have ended me? What if it hadn't?

So with my second child, Edwin, I thought we could do things differently. His labour was unexpected, which seems like a ridiculous thing to say about a state literally called 'expecting', but he was nine days early. It was awful timing: my mother, who was travelling from Texas to take care of our now two-and-a-half-year-old son, wasn't due for another two days. We were expecting a new mattress to arrive in three days. We had a grocery delivery coming in the morning. I had an article due the next day. And it was 4 a.m. But time, tides and babies wait for no (wo)man.

I wasn't quite dilated enough when we arrived at the hospital, so we waited for my sluggish cervix in the birthing centre's waiting lounge. Our toddler, still in his brown and green striped pyjamas, was ecstatic. It was 5.30 a.m. He'd just got to ride in a taxi, there were mats to roll around on, giant yoga balls to bounce on. He found a paper barf bucket, put it on his head and declared, 'Look at me, I'm a farmer!' When I tried to get on all fours in cat's pose to ease my aching back, he wedged himself between my giant belly and the mat and made faces up at me. When I muttered 'Oh my god oh my god' during contractions, he grinned and muttered 'Oh my god' too.

I couldn't stop laughing, even through the contractions. Nevertheless, we recognised that this whole birth thing might take a while and two-and-a-half-year-olds aren't known for their patience. We rang a very good friend to come and pick him up for a day of brunch and cartoons.

The lounge was suddenly a lot quieter. I was suddenly in a lot more pain. But when the admitting nurse came in to ask whether I'd like to stay in the birthing centre or head down to the ward – and the possibility of an epidural – I surprised everyone, including myself, by saying I wanted to stay. My husband asked the nurse to give us a minute. 'Are you sure?' he asked, worry plain on his face. He'd been there the first time around, of course, and he remembered what happened when the contractions got worse. But I was sure.

It's hard to recapture exactly what I was thinking at that point; memories are generative, perpetually reshaped and coloured by everything that has happened since, by what I know now and what I wanted to be true. But in my recollection, I was curious. I'd done the work of the labour – my destroyed pelvic floor could attest to that – but I still didn't know what it actually felt like to push a baby out of my body. I wanted to find out what was on the other side of that terrible, burning wall of pain, to find out if I could get there and hold myself together in that pain. I was also relaxed in a way that I hadn't been the first time around; small children are brilliant at supplanting your personal needs, and that meant that while our toddler was there, I wasn't stuck in myself. Even though he'd gone now, I was still lifted by that. So yes, I was sure. Mostly.

On the way to the room where I'd give birth, we rescheduled the grocery order for the next day and discussed using the birthing pool, or the Hot Tub Baby Machine as we insisted on calling it. Sure, I said, let's do it. It felt like I was on a roller coaster, listening to the *tick-tick-tick* as the car climbed the first hill. There was no getting out now.

Over the next four hours, I lay on a hospital bed and breathed nitrous oxide to take the edge off the contractions; it tasted sweet and evidently made me a charming conversationalist. When it hurt – and it did, a lot – I breathed. The midwife team filled the tub. I ate a Snickers bar. The contractions had long since reached that feverish, burning stage, but something about having made the decision to do this meant I wasn't panicking. This was my choice, I was on the roller coaster and I was going to ride this bastard down. Just hand me the nitrous.

It wasn't until Edwin was nearly crowning that I was allowed to finally get into the tub; the midwife team had worried that the transition to the tub would slow labour down. This meant I had to waddle half-naked across a wet floor and up two little stairs with a fully engaged baby between my legs to climb into a tub. But when I got in, they were right – the sense of immediate weightlessness eased the pain in my back and everything seemed to slow down.

After about a half-hour of pushing, Edwin's head started to emerge. I was too tired and too scared to pull him out myself, so the midwife reached in and grabbed him, bringing him to the surface of the water and placing him on my chest. He was as blue as a Smurf. 'That's totally OK,' she said. 'Are you sure? Are you sure that's OK? Is he OK?' my husband and I both gabbled. But he was already starting to pink up, his beautiful eyes open to the world. I didn't hurt quite as much any more, or rather, if I did, I didn't notice. We went home that night – after all, we had a grocery delivery coming the next day – and I felt good. Strong. Ready.

During my first labour, I was nervous. It wasn't the pain that floored me as much as it was the fear of the pain. When it began to reveal itself in full, I ran for the epidural because I had no idea how bad it would be and I didn't want to find out. My fear of the pain, the panic it induced, was in many ways worse than the sensation itself. But I lived through it; I also learned that my fear of the pain was nowhere near the worst thing about becoming a mother, a process that is wonderful and terrible and feels like you've grown a new limb, with new nerves that jangle at the slightest touch. The second time around, I felt ready to meet that pain.

For years after both my birth experiences, I thought of how I survived them as my secret superpower. I had already done something hard, something painful, and lived. If something hard and painful came up again, well, it probably couldn't be worse than that. What these experiences gifted me was the knowledge that pain isn't an absolute. It's negotiable. Sometimes.

We need pain

Our experiences, our stories, shaped our relationships with pain. But they also raised a bunch of questions for us. How and why is pain sometimes negotiable? What makes one experience empowering and the other traumatic? If we could just make pain go away, shouldn't we? What is the purpose of pain? And how can we ever approach pain as anything but negative when our body is screaming or in the wake of so much *hurt*?

We decided to think about pain using tools of our respective trades – lots of research and putting these questions to the right people. In both of our disciplines, sociology and journalism, we talk about the practice of 'making the familiar strange'. Essentially, this is the idea of taking something that appears to be simple and obvious – something you might say, 'Well, it just *is*' about – and asking 'But *why* is it?' and '*What* is it?'

The people we asked included neuroscientists studying how pain is made in the youngest, most vulnerable populations; endurance athletes who run further in one day than most people drive in a week; psychotherapists helping people rethink their chronic pain; cardiologists teaching angina patients to conquer their heart pain; BDSM aficionados who push the boundaries of sensation; and mothers who've experienced orgasms at birth. We've talked to people who use pain as performance art, and those who use pain in the practice of sacred ritual. We've met people who live with pain every day, and people who have never – not once – experienced pain.

What we learned is that pain is so much more than bad. Pain is necessary, providing us with vital information: that thing is sharp, this thing is hot, gravity isn't always your friend. Pain is useful and not only as real-time feedback about what's happening in our bodies. Pain can be a tool to regulate our emotions; it's why we go for long runs or jump in icy lakes. It can settle us in the moment and remind us that we're still here, still alive. Shared pain can offer a pathway to empathy and connection, to spiritual enlightenment, to salvation. And pain itself can be rewarding – research has long demonstrated that pain sensitises us to our sensory experiences; experiencing acute pain can make food taste better and touch feel more intense in the immediate aftermath. In the right context, pain can feel good.

The stories we've heard demonstrate that we need pain as part of a large, diverse emotional vocabulary. Across these conversations, investigations and experiences, we were frequently reminded that sometimes pain is simply the worst. But regarding all pain as pathological means we regard it as static, immutable; we miss out on opportunities to transform that experience or even to gain from it. The stories we've heard also underscore the fact that medical practitioners cannot be the only authority on pain – a medical belief system that focuses on suppression and avoidance of pain, that defines pain only as a *bad thing*, doesn't just strip pain of its useful qualities it also wrests control of that experience from the individual. We, the patients, must educate ourselves to more fully understand how our bodies and brains work and, through that, learn to have a better relationship with pain. Pain is fundamental to human experience. It's not the enemy.

This book is about how pain is made, how our brains and bodies create this experience. Once we have a better understanding of how pain is made, we can remake it.

CHAPTER ONE

What is Pain?

Steven Pete had regrets from the moment his bike launched into the air, with him on it. Thinking, 'Oh man, this is not going to be good,' just before he plummeted, bricklike, to the ground many feet below. He was right: it wasn't good.

The idea had been to ride his bike down the very steep hill, then thunder up the ramp, a piece of plywood propped against a guardrail, and soar, like Evil Knievel, over the deep ravine to land the other side. He was 11 or 12, as were his friends, so any concerns were brushed aside by the promise that this was going to be 'cool'. He mounted his bike at the top of the hill, steadied himself and took off.

It wasn't hitting the ramp that was the problem so much as everything that came after that. After catching more air than anyone expected, Pete and his bike began to lose it just as quickly, an eventuality no one had really considered. They both landed in a crumpled heap at the bottom of the ravine. Winded, Pete stood up, dusted himself off and started the scramble back up the bank, trying to drag his bike behind him. Before he'd made it back up the side of the ravine, his right leg had swollen like a cooked sausage from his ankle to his knee.

'Everyone kind of freaked out,' he recalled, now 38 and a lot wiser. His friend's parents called his parents, and they immediately took him to the local hospital near his hometown in Washington state, near the Oregon border.

Pete didn't freak out, however. He knew he'd injured himself – that was obvious from the sensation of throbbing and from the fact that his leg didn't seem to want to support his weight any more. But it didn't hurt. Where most people would have been in agony, Pete wasn't. In fact, he had never experienced pain. Not once, not in the dozens of times he'd visited this same hospital with broken bones, cuts and burns, not *ever.*

It might seem counter-intuitive to start a book about pain with the story of someone who has never experienced it. But Pete's condition demonstrates a couple of really important points about how pain works and why we need it, about what pain is and what it isn't. Pete was born with congenital insensitivity to pain, an extremely rare genetic condition that means he can't feel pain or, more specifically, that he can't perceive nociceptive stimuli. This distinction is super-important and, believe us, we'll be discussing it at length later. Suffice to say that if you took a knife and sliced Pete's arm, he'd certainly look at the growing red line with alarm, but it wouldn't be because it hurt.

Congenital insensitivity to pain, or CIP, affects just a handful of people across the world; fewer than 30 people in the UK have been diagnosed with it, and there are maybe a few hundred people worldwide living with it. Researchers put the chances of having CIP at around one in a million. It is genetic – Peter's brother, Chris, who was less than a year younger than him, also had it; he died by suicide in 2008 after doctors told him he'd be confined to a wheelchair for the rest of his life, in that it affects the function of the peripheral nervous system. Pete can't smell anything either, a condition called anosmia. CIP can also come with anhidrosis, the inability to sweat normally and therefore to thermoregulate the body, and developmental delays, neither of which affected Pete or his brother. CIP, of course, presents enough of its own challenges and by that time in his life, Pete was well known to the staff at the hospital. 'At that age, I'd spent the majority of my childhood at the hospital,' he said. 'Whenever I'd go in there, everyone knew who I was, they knew my condition.'

Pete's parents first realised his pain responses were different to other children when he started teething and chewed off a large chunk of his own tongue. 'That was pretty terrifying for them,' he said, which seems like a stunning understatement. They brought him to his paediatrician, who had luckily just read about another case of inability to perceive pain. He held a lighter to infant Pete's foot. When Pete still hadn't reacted to the heat of the flame after his sole started to blister, the doctor pronounced him unable to feel pain and suggested his family take him for further tests at a larger hospital in Seattle. There, he was subjected to another round of 'noxious stimuli', this time a small rolling wheel covered in spikes ('Pretty much the same sort of test but done differently,' he said). Again, little Pete exhibited no reaction. The diagnosis was CIP.

And that was it. This was in the early 1980s, but even so, the medical reaction to Pete's condition was muted – no university research labs clamoured for his blood, no case studies appeared in major journals, no write-ups in magazines, nothing. Just a warning to his parents that they'd need to watch him carefully: people with his condition sometimes don't know that they've injured themselves until it's too late. They also tend to be prone to infections, not only because they're unaware that they've broken a bone or cut themselves, but also because they can't feel the pain that typically results from an infection; there is also some evidence that people with the disorder exhibit a reduced immunity to *Staphylococcus aureus* bacteria.

That inability to perceive tissue damage and tendency for infection means that the life expectancy for people with the condition is significantly shortened; some figures say half of children with CIP die before the age of three. The spectre of an early death has loomed over Pete since he was a child. 'I remember a lot of times growing up, doctors saying, "People with your condition don't live past X years old,"' he recalled. 'And then, "People with your condition don't live to be teenagers." And then when I was a teenager, they'd say, "People with your condition don't live to be young adults." You're constantly, always fearing, thinking, "Damn, dude, my mortality rate is so short compared to everyone else, and I might as well go out and do whatever I want."' For a little while, he did – letting friends ride their bikes over him in the driveway, taking punches because he could, pulling Evil Knievel stunts because they were cool. But pain exists for a reason and people who don't feel pain are typically held up as proof that we need pain to survive, and they' re good evidence – from an evolutionary perspective, Pete shouldn't be alive.

At its most basic, pain is a warning system, letting us know that tissue damage is imminent or is occurring. If we are injured, the injured area continues to hurt to ensure that we rest and allow the tissue, bone and muscle time to heal. If there is an infection present in our bodies, it hurts, alerting us to take action. This efficient system evolved in an environment where dangers and potential injuries lurked pretty much everywhere, and it has kept us alive for millennia.

What's kept Pete alive without it is vigilance. His family were forced to turn what should have been a personal, automatic, unconscious mechanism into a very conscious one. Which was exhausting and difficult, to say the least. When Pete was just a toddler, his frequent visits to the hospital alarmed child protection authorities in Washington state and he was briefly removed from his family's home. While under observation in a Seattle hospital, he broke his foot. Those charged with watching him didn't notice it until more than a day after the break, and they never figured out how he'd done it. 'Child Services said they realised what we were going through,' Pete's mother, Janette, later told a local newspaper in 2010. When he was five years old, he broke his arm; at the follow-up appointment, the orthopaedic doctor scanned the x-rays and gave Pete the all-clear. His mother, however, didn't agree – the arm smelled, she said. A trip to the hospital confirmed that the bone was infected. Pete lost part of his bone and spent the next two weeks on intravenous antibiotics.

As Pete got older, he wasn't always as vigilant as he could have been – often he wouldn't be aware of an injury right away, in part because 'every injury has its own distinct feeling to it,' he said. Once, he visited his doctor because the movement in his arm and shoulder was 'off', while his back also felt 'weird'. Turns out, he'd fractured three vertebrae in a tobogganing accident eight months earlier.

Pete is a lot more careful with himself now; tobogganing is out. His last broken bone was 13 years ago. 'I've adapted to it, I know it,' he said. 'You just take all these past experiences and you kind of learn something new from each one ... You get to this point that you're like, "OK, I know that this action is going to cause this to happen, so I'm just going to avoid doing that." You learn.'*

* Pete also 'learned' to smell, or rather, to feign the ability: 'When you're little, a lot of your behaviours are learned behaviours, you kind of mimic what other people do. So if someone says, "Oh, I just farted, it really stinks," you start to do what everyone else is doing, going, "Oh, god, uughh..."', he said, laughing. 'But by the time I became a teenager, I realised, yeah, I couldn't smell.'

Sounding the alarm

Pete had to learn to take care of himself because he was born without a warning system; we'll come back to why in a bit. But first, we need to really get into how it works in most people – people who are not Pete.

Say you've just dropped a 15-pound frozen turkey on your big toe wrestling it out of the freezer. The event sets in motion a chain of relayed signals: first, your nociceptors, the specialised sensory neurons that report damage, send a swiftly moving signal from the crushed toe up nerve tracts in your leg to a relay centre in your spinal column. From there, the signal is conveyed up to your brain and then projected out through a vast network of neurons and glial cells, generating an immediate reaction – your heart rate increases, your breathing accelerates, you scream a few choice words. All non-essential neural processing is suspended briefly as attention and resources shift to protection and self-care.

While you're hobbling around your kitchen, cursing that stupid bird as well as the holiday that occasioned its existence, yelling for someone to *please* bring you an ice pack, your nervous system is taking further action. Your system kicks into high gear, deploying your endogenous painkillers (the family of pain-suppressing chemicals that make up our innate opioid and endocannabinoid system) to dull the pain while immune-boosting and infection-fighting white blood cells move to the injury site and start to repair the damage. The initial shock gives way to a dull throbbing, your toe starts to swell and turn exciting colours, and the ache that sets in is your body saying get off your foot and maybe do takeaway this year.

That's the basic roadmap of what happens when something most people would consider painful unexpectedly occurs, but the devil is in the details and it's worth looking in more depth at exactly how this very efficient system accomplishes all that.

It all starts with the nervous system, which is tasked with keeping us alive. The nervous system regulates everything from the skin inwards and attempts to control as much as possible from the skin outwards. Structurally, it is organised into subsystems according to purpose and location; each of these is involved in the generation and perception of pain, as well as our response to that experience. The central nervous system (CNS) includes the brain, brain stem and spinal cord, and is responsible for, among other things, processing sensory data. The peripheral nervous system (PNS) contains all the nerves that lie outside the brain and spinal cord, describing a complex blueprint of information-carrying nerve pathways between the skin, internal organs, tissue, muscle and the CNS.

Under the PNS umbrella, the somatic nervous system is responsible for voluntary control of the body and relies on sensory nerves that relay information from the periphery to the CNS and motor nerves that carry messages from the CNS to the periphery. Also comprising the PNS are the systems that function without our conscious awareness, called our autonomic nervous system. This branch is further subdivided into the enteric or intrinsic nervous system, which controls our gastrointestinal activities, and the sympathetic and parasympathetic systems. The sympathetic nervous system prepares us for stress, anything that is going to tax our body's resources; this is the network responsible for 'fight or flight' or, more accurately, the threat response. The parasympathetic nervous system is commonly referred to as the 'rest and digest' system, or the 'you need to calm down' system, but it also plays a big part in sexual arousal, the 'freeze' response to threat and a host of other physiological functions.*

This interconnected network communicates via nerve impulses conveyed along pathways made of the primary cells of the nervous system, neurons and glial cells. Neurons come in all shapes and sizes and, like other cells in the body, they have a nucleus containing genes and organelles, and engage in cell behaviour, including protein synthesis. But unlike other cells, neurons have processing arms, called axons and dendrites, which extend from the cell body and form

---

* Herophilus of Chalcedon, a physician-philosopher and pioneer in the use of dissection in medical study who lived in fabled Alexandria in the third century BC, was the first to demonstrate that the nervous system was a separate entity from the cardiovascular system, and to note that there were sensory and motor neurons. He was one of the first to fix the location of human action and intention in the brain, explaining that the brain was also responsible for the activation of the systems in the body; even more remarkably, he distinguished between the cerebrum and the cerebellum in the brain. And he may have learned all this by practising vivisection on unfortunate human subjects.

connections with other neurons. Neurons typically – but not always – have one axon, which is a cable-like construction joined to the neuron at a junction called the axon hillock. These can extend far from the cell body – motor neurons at the base of the spinal cord have axons that stretch as long as three feet.* The axons of many neurons are bundled together, forming the ropelike constructions better known as nerve tracts. Dendrites are the spiky branch-like structures off the main cell body; they extend outwards, reaching for information in the immediate vicinity of the neuron. The junctions between two neurons – well, between their axons and dendrites – are called synapses. Each neuron can have upwards of 10,000 synaptic connections to other neurons and to non-neuronal elements, such as skeletal muscle fibres and hormone-producing glands like the pancreas and ovaries.

Primary afferent neurons that send information to the CNS can be grouped by size and are, from largest to smallest, A-alpha, A-beta, A-delta and C-nerve fibres. The thickness of the fibre determines how quickly the signal can travel – the thicker the fibre, the faster the signal. And signals can travel *fast*: impulses from the spinal cord to our muscles, for example, move at about 270 miles per hour. To do this, neurons need the help of the equally important and most abundant type of cell in the nervous system: glial cells. These provide structural support, including producing the myelin sheath, a fatty substance that wraps around the length of some axons and helps the current glide along.

Neurons communicate through electrical and chemical signalling, using neurotransmitters. These are uniquely designed endogenous chemicals that bind with specialised receptors located on a neuron's membrane that are responsive to specific types of activities and stimuli.

---

* We know how axons work because of squids. Or, more precisely, the squid's giant axon. Your basic squid possesses a neuron with a giant axon 1mm in diameter that extends all the way through its tail. This axon controls the squid's water-jet propulsion system and, helpfully, is large enough to stick a very fine micropipette down. In 1939, University of Cambridge researchers Alan Hodgkin and Andrew Huxley did precisely that: they inserted a thin, saline-filled glass tube containing a chlorided silver wire down a squid's giant axon, to act as a non-polarisable electrode. Through a series of experiments, they were able to record the potential difference between the interior of the fibre and the exterior, effectively demonstrating action potential at work. In 1963, they won the Nobel Prize in Physiology or Medicine for their theory of nervous conduction.

They work like one of those puzzles for small children, where you push the shape through the correctly shaped hole – the star through the star hole, the oval through the oval hole, that sort of thing. But, as parents also know, more than just the plastic shapes that came in the box can fit through those holes, for example, noodles that suggest the shape of stars can also, apparently, be a fine fit. We'll talk about this more later, but one way analgesics work is because they are similar to our home-grown neurotransmitters.

Without going too far down the chemistry rabbit hole, neuronal communication works like this: when a neuron is not signalling, its electric charge relative to the area outside the neuron's semi-permeable membrane is negative. The introduction of a stimulus, acknowledged by a necessary receptor, opens up ion channels on the neuron, allowing positively or negatively charged chemical ions to move through the neuron's membrane. This depolarising current triggers an explosion of electrical activity, called an 'action potential', inside the neuron. If the neuron's interior charge drops to around -55mV – its firing threshold – then the action potential is released and the message is deployed. Neurons either fire or they don't, and there is no difference in degree. If a stimulus is registered more intensely, it's because more neurons are firing.

Once deployed, the action potential moves down the axon to its synaptic terminal, where the impulse triggers the migration of vesicles, little packets that contain neurotransmitters, towards the adjoining neuron's presynaptic membrane. There the vesicle fuses, releasing neurotransmitters into the synaptic cleft, where they diffuse and bind to their uniquely designed receptors on the dendrites of surrounding neurons, opening their ion channels and causing a depolarising event in that neuron. And so it goes, from axon to neighbouring dendrite, dendrite to the cell body, cell body to axon and on to the next.

Our billions of neurons* are organised into enormous, complex interconnected networks; scientists estimate that our neuronal

* Exactly how many neurons do we have? Good question. Counting neurons is a contentious business; one research team that did in 2009 determined there were around 86 billion neurons in the human brain alone. However, another review, published in 2015, suggested a revised number could be between 40 and 136 billion. So the answer is a lot.

network is made up of between 100 and 1,000 trillion synaptic connections. Communication between this inconceivably vast horde of neurons is what allows us to function; the interactions of these little guys define who we are. At any given moment, there are billions of signals being sent, received and interpreted in our nervous systems, billions of conversations happening all at once. So which conversations are about pain? How does an electrical impulse become a perception of pain? And what happens when signals fail – when they get lost or garbled, or start blaring?

Making signals into sense

In December 1895, German engineer and physicist Wilhelm Conrad Röntgen was experimenting with passing electrical current through low-pressure gases when he started to observe some unexpected outcomes. Specifically, he found that objects caught in the path of his discharge were captured in varying degrees of transparency on photographic plates. The next step was to drag his wife, Anna Bertha, into the picture – literally. The image he captured of her hand, held in the Wellcome Library's collection, shows her bones and wedding ring in faded sepia brown, while her flesh simply isn't there. This was the first x-ray image. The ghostly image terrified Anna Bertha, who reportedly gasped, 'I have seen my death!' She wasn't exactly wrong.

Röntgen mailed copies of his images and a draft of his paper on his discovery to fellow physicists across Europe. Word spread quickly; within weeks, other scientists – and amateurs – were replicating Röntgen's process. Doctors began using their home-made machines to detect breaks in bones and the location of bullets in bodies, engineers were pointing 'Röntgen's rays' at anything that would stand still long enough, and entrepreneurs were figuring out how to capitalise on the novelty – in Chicago, 'x-ray slot machines' allowed customers to see the bones in their hands for a coin.

Soon, however, reports started circulating about the more insalubrious consequences of exposure to Röntgen's process. X-rays are, after all, high-energy electromagnetic radiation. People who used the rays to take images of the head noticed hair loss in the irradiated area; others reported skin blistering, bloodshot eyes and swollen lips. In December 1896, a woman reporting abdominal pain underwent three x-rays in a single day. Two days later, the skin on her stomach sloughed off. By July 1897, the *New York Times* noted that courts had awarded several individuals 'heavy damages' after they'd suffered 'injury and disfigurement' after an x-ray. Then came the deaths: in 1904, Clarence Dally, Thomas Edison's chief x-ray technician, became the earliest reported death linked to x-ray use – he died of metastatic carcinoma at the age of 39. More deaths followed, mostly of aggressive cancers; Röntgen, who'd won the Nobel Prize in 1901 for his discovery, died of colorectal cancer in 1923.

Röntgen's incredible rays demonstrate an essential point: our perception of sensation is limited to the types of receptors our somatosensory neurons come equipped with. The big five senses – taste, smell, sight, hearing and touch – as well as our sense of balance and movement are thanks to the specialised receptors located on neurons that are designed to 'play' the sensations we've come to know. Early x-ray users didn't know that their exposure to radiation was killing them because they – and we – don't possess receptors that are able to perceive radiation damage as it's happening. (This is the point where the *Twilight Zone* music plays and you stop and wonder what else is out there that we can't perceive.)

We do, however, have photoreceptors to process light; mechanoreceptors to sense pressure and vibration; chemoreceptors that are sensitive to specific chemicals; thermoreceptors for temperature; and while we may often wish we didn't, nociceptors to process tissue damage or the potential for it.* As Röntgen's rays revealed, our bodies don't come equipped to sense all potential threats, but we do have nociceptors that are sensitive to the chemical by-products of cell destruction, which is why victims of radiation poisoning felt pain as

* Unsurprisingly, other species developed receptors that responded to their unique needs and environments; insects, for example, have hygroreceptors, which provide information about the moisture content of their environment. Some species, meanwhile, lack certain receptors – the naked mole rat, for instance, is impervious to certain types of noxious stimuli, specifically acid and capsaicin (the active component in chilli peppers), because it lacks the specific chemoreceptors and mechanoreceptors to sense them.

skin cells began to die off. We also have mechanical nociceptors that are highly responsive to force and pressure damage, such as a frozen turkey on the toe or a hammer meeting a thumb. These are also the receptors that report on cuts and incisions. Chemical nociceptors are highly responsive to a wide range of irritants, such as strong acids, toxins in the environment and the venom of snakes or insects. Thermal nociceptors are responsive to damaging hot and cold temperatures.

Sleeping or silent nociceptors, found in our skin, organs and deep tissue, are 'awakened' once damage has occurred and inflammation sets in. These nociceptors are crucial after injury has occurred; they are involved in hyperalgesia, the increased sensitivity to noxious stimuli, and allodynia (when typically non-noxious stimuli, such as a gentle touch, for example, is experienced as pain). Both are normal, necessary responses to injury; part of the reason we don't walk on a broken toe, giving it time to heal, is because it hurts.

Our sensory neurons aren't only specialised to specific stimuli but also to specific ranges or intensities of those stimuli. For example, we have receptors in our cochlear ducts, within the cochlea inside our ears, that transform vibrations in the air into meaningful sound. But we humans are only able to hear frequencies that range from about 20 to 20,000 Hz,* our adequate stimulus range. These thresholds vary by person and can change with experience and age, a point that inspired the high-frequency ringtone only heard by those under the age of about 40. Other species may have similar auditory mechanisms but different ranges of adequate stimuli; dogs famously can hear between 67 and 45,000 Hz, but moths blow everyone away: they can hear up to 300,000 Hz. In nociception, for instance, this means there are certain types of thermal nociceptors that respond only to temperatures greater than 45°C and lower than 15°C, a kinds of human danger zones.

---

* Other receptors can still sense the vibrations, however. For example, though our auditory cortex isn't registering it, we can sense infrasound, vibrations lower than 20 Hz. Infrasound might be one explanation for the feeling of being haunted – according to some research, sensing infrasound can induce a general unease, while the vibration of the eye can result in seeing things that aren't there. Perfect ingredients for a haunting.

A specialised nociceptor is responsive to specific types of stimuli, and like other sensory neurons, they only have one output modality – meaning that, no matter what initiated the firing, it can only play one note. For example, menthol uses the same ion channel as our cold thermal receptors, which is why we get the pleasant sensation of coolness with our mentholated toothpaste, body wash and cigarettes. This 'one note' situation is why intensely hot or cold temperatures both feel like burning (explaining why people experiencing hypothermia feel the urge to strip off all their clothes). This also explains why chilli peppers are 'hot' – the active ingredient in peppers, capsaicin, is structurally similar to other noxious stimuli, inducing the same burning sensation. Fun fact: foods that are high in capsaicin burn going in *and* coming out because we also have receptors that register capsaicin in our anus. Even funner fact: capsaicin is one of the weird ways that noxious stimuli can actually relieve pain.

Fascinatingly, not all body parts are equipped with nociceptors, meaning that while they can definitely be damaged, they can't 'hurt'. Though it's gruesome to think about, you wouldn't experience pain if a surgeon were to slice into your spleen. (Cutting through the layers of dermis and muscle to reach your spleen, however, is a whole different ball game.) The key to knowing which organs we can 'sense' and which we can't has to do with exposure to the outside world and our ability to act in it. Consider the organs that do generate sensations that reach our conscious awareness: the stomach, bladder, trachea, uterus, the vagina and penis. These 'hollow' organs (meaning if you were to cut into them things would fall out, liquid or otherwise) are all involved in processing activity coming from the external world, in addition to processing changes arising from internal activity. It makes sense that they would come with nociceptors, because potential threats typically come from outside. However, just because some parts of our body do not have a 'direct line', so to speak, it doesn't mean they don't have ways of getting the signal out. Organs that don't have nociceptors to communicate tissue distress can 'refer' the pain to somewhere that can get a connection, and sometimes that connection is nowhere near the site of injury. This is frequently the case with visceral pain – the pain emanating from internal organs – which is not nearly so clear-cut as, well, a clear cut.

Specialisation of receptors is one of the later systems to develop in the gestating human. Thanks to recent advances in imaging technology and the consenting parents of preterm infants, studies consistently demonstrate that prior to 35 weeks of gestation, noxious, tactile, visual and auditory stimulation of a premature infant produces non-specific brain activity that researchers call a 'neuronal burst'. Essentially, this brain activity signals that *something* sensory is happening but not what. However, between 35 and 37 weeks of gestation, sensory and nociceptive specific brain activity emerges – the soft stroke of a cotton swab generates different neuronal firings to the pierce of a first vaccination.

This implies that our nervous system has evolved to distinguish and inform on different types of sensations because it was important to our survival. But even after birth, our nervous system is not done developing; we have to learn what these signals mean and what to do with them. In order for our CNS to develop adaptive responses to nociception – meaning we produce the right amount of endogenous painkillers in response to or anticipation of injury – we have to have opportunities that require both activation and inhibition of our sensory network. In other words, we need to train ourselves.

Think of our first years of life as a series of calibration tasks. It starts with stimulation: the gentle touch of a parent's hand across an infant's cheek, the warmth of skin-to-skin contact and the satisfaction of feeding, or the chill of an antiseptic wipe and the sharp sting of a needle prick. Over time, a baby's developing circuitry forms patterns on how different sensations pair together, becoming predictable, allowing for faster and more efficient processing. This refinement is illustrated in recent work by researchers at University of Oxford that recorded both electromyogram (EMG, which measures motor neuron activity) and electroencephalogram (EEG, which measures electrical activity in the brain) during tactile and painful stimulation in infants between 28 and 42 weeks old. They found older infants showed more nociceptive specific brain activity and less withdrawal reflex compared to the younger infants. This suggests that as our CNS gets better at distinguishing and processing pain, our reflexes become more efficient and essentially 'calm down'.

Wherever we are in our neural development, all sensory signals from the periphery – whether that's cold or heat, pain, touch, tickle* or itch† – are relayed through a chain of three neurons; we alluded to this earlier in our frozen turkey example. A sensory neuron fires and the signal is relayed to the second neuron via a synapse located in the dorsal horn at the back of the spinal cord. This synapse adjoins a group of neurons called 'wide dynamic range' neurons, which receive inputs from the skin, muscles, joints and organs, and respond to primarily tactile stimulation of all kinds. From here, the signal travels up the spinal cord to the next neuron in the chain, located in the thalamus, a kind of relay station with extensive nerve connections and composed of grey matter located just above the brain stem between the cerebral cortex and the midbrain. It's the thalamus's job to broadcast the message about the injury that has just happened, in order to initiate a response. (In the case of the frozen turkey and the broken toe, the appropriate response was gin.)

---

* Aristotle believed that only humans were susceptible to being tickled, owing to the 'fineness of their skin' and their 'being the only creatures that laugh'. He offered this as further evidence of human superiority and so would likely be sad to learn that we are *not* the only species who are ticklish: rats are also ticklish, and their so-called *freudensprünge*, or 'joy jumps', suggest they also quite enjoy it. In rats, at least, evidence suggests that tickling is 'fun' because it initiates play fighting, which in turn is important for healthy development. There are two types of tickle, according to psychologists G. Stanley Hall and Arthur Allin, writing in 1897: gargalesis, which is the rougher, laughter-inducing tickling, and knismesis, which is the light-touch tickly-itch. And you can't tickle yourself because, well, you know you're going to tickle yourself.

† Itch, or pruritus, is a sensory experience related to but distinctly different from nociception. Basically, pruriceptors are specialised itch receptors that work a lot like nociceptors, in that they can respond to mechanical, thermal or chemical stimulus, but are carried by C-fibres. Scratching an itch activates other receptors, including nociceptors, that interrupt the itch signal, but only for a time.

Weird wiring

Warning: we're about to pull the rug out. Up until now, we've made the whole process of sensory experience sound fairly straightforward, even setting up Steven Pete's malfunction of the nociceptive system as a weirdest case scenario. His case *is* unusual, but this doesn't mean so-called 'normal' systems are any less complicated. A lot – *a lot* – happens between the prick of a needle and the '*Ouch!*'

For one thing, it's hard to express just how much continuous intrinsic activity, how much *noise*, there is going on in our nervous systems at any given moment. With all the chatter, our networks are also tasked with prioritising signals, amplifying important ones while ignoring others. In the 1960s, Canadian psychologist Ronald Melzack and British physiologist Patrick Wall's research demonstrated just this, and further offered a theory as to how we might intentionally manipulate the nociceptive signal to manage pain. Melzack and Wall's gate control theory proposed that we could block nociceptive signals by stimulating other tactile sensory receptors. Tactile sensations travel faster via the larger A-alpha and A-beta nerve fibres so, according to their theory, these signals could arrive first and 'block' the nociceptive signals arriving via the smaller, slower A-delta and C-fibres. Their theory seemed to test out: applying transcutaneous electrical nerve stimulation (TENS) via electrodes in positions above the painful region did result in lower reports of pain. Further studies revealed it's not quite as simple as 'blocking the gate' of pain signals – scientists now believe the relief from using TENS units is a result of counter-irritation – but Melzack and Wall's gate control theory of pain brought much-needed attention to the fact that somatosensory signalling is nowhere near as linear as we'd like to think.

For another thing, when things go wrong with nociceptive processing, even just a little bit, the consequences can be enormous in terms of how we perceive sensory input. The fault in Steven Pete's wiring is down to how his neurons communicate – or, rather, how they don't. The error wasn't entirely clear until 2012, when genetic testing revealed that Pete has an autosomal recessive mutation on the SCN9A gene. This gene, also known as 'sodium voltage-gated channel alpha subunit 9', belongs to a family of genes responsible for providing the instructions for making sodium channels, which transport positively charged sodium ions across cell membranes to trigger the action potential. The specific gene encodes the alpha subunit of a sodium channel called NaV1.7; these channels are found in nociceptive nerve cells as well as in olfactory sensory neurons, explaining the anosmia. Pete's NaV1.7 channels, in basic terms, don't work, so his sensory neurons lack the ability to 'play' nociceptive and olfactory stimuli.

Had Pete's mutation happened in just a slightly different way, he could have been dealing with another pain condition. Mutations impacting this same sodium channel can also cause erythromelalgia, a condition that results in painful swelling and inflammation of parts of the body, usually the hands or feet; paroxysmal extreme pain disorder, which triggers attacks of severe pain accompanied by flushing of the skin and elevated heart rate; and small fibre neuropathy, which produces attacks of pain and an inability to discern hot and cold. In all of those conditions, the mutation essentially leaves the sodium ion door open and the pain signal perpetually blaring.

If one mutation on a sodium ion channel can leave an individual either insensitive to pain or constantly plagued by it, just imagine what other problems in the wiring can do. For example, familial dysautonomia is a genetic disorder that, though very rare in the general population, affects around 1 in 3,700 people of Ashkenazi Jewish descent. It is caused by a mutation on the ELP1 gene, which provides the instructions for making a protein called the 'elongator complex protein 1', part of a pack of six proteins used in cells throughout the body. Those with the mutation don't make enough of the protein, which in turn affects the development of certain nerve cells in the autonomic nervous system and in the sensory nervous system, resulting in, among other symptoms, reduced sensitivity to pain and temperature changes. But exactly *why* deficiencies of this protein result in pain insensitivity isn't clear.

Other faults in the system can occur through disease or damage. Leprosy – Hansen's disease – causes localised insensitivity to pain owing to the damage to peripheral nerves; long-term infection of the *Mycobacterium leprae* bacteria may cause degeneration of the myelin sheath, inhibiting the neuron's ability to communicate. Now, you'd think that being able to identify the fault would help us to fix it, but the reality is that the complexity of these systems routinely defeats us. Degeneration of the myelin sheath might be part of what reduces pain sensitivity in people suffering from leprosy, but it's also part of what makes multiple sclerosis *more* painful. Multiple sclerosis – an autoimmune condition with an unknown cause – attacks the myelin sheath, causing numbness, disrupting coordination and contributing to neuropathic pain.

Even hypertension, a much more common condition, is known to produce reduced sensitivity to acute pain, although the exact mechanism behind this is unclear; some researchers have suggested that the relationship between the cardiovascular system and pain regulatory systems may be compromised. Adding to the mystery, however, is the fact that elevated blood pressure in patients with chronic pain is associated with *increased* sensitivity to acute pain and more intense chronic pain.

This extreme variability in how nociceptive signals are conveyed underscores the challenge in treating chronic pain, which affects more than 20 per cent of the global population. Chronic pain is loosely defined as pain that lasts longer than three to six months, roughly the time that it takes for tissue damage, even the most severe, to heal. In some cases of chronic pain, there is a more obvious reason for a persistent signalling error. Nerves can be damaged by acute injury or through disease, and this damage can happen at any location, from the nerve fibres in the tips of our fingers to the deepest parts of our brain, with serious implications for how we perceive sensations. For example, central neuropathic pain, which can create sensory 'hallucinations' such as feeling intense cold or hot, can result from injury to the thalamus, the sensory information relay station deep in the brain.

And when things are damaged, they very seldom can be restored to good as new. Take our skin – when it is cut or burned, it heals with tissue made of the same stuff, collagen. But the highly efficient, random basket-weave of collagen fibres that we had before is now replaced with a lower-quality weave we call scar tissue, which is less flexible and less resistant to ultraviolet radiation. It gets the job done but not quite as well.

Chronic pain might be a similar situation. Researchers haven't uncovered all the precise mechanisms, but the idea is that chronic pain can result not just from injury but also from dysfunction in the recovery process. For instance, complex regional pain syndrome (CRPS) is a post-traumatic chronic condition in which pain persists, typically in a limb, long after an injury. A complete pathophysiology is still some way off, but one suggestion is that CRPS can result from altered immune system response. Research has found that pathological autoantibodies – antibodies produced by the immune system against a person's own proteins – are present in people with persistent CRPS; the more autoantibodies, the more severe the reported pain. Why those autoantibodies are there in the first place isn't entirely clear, however, other researchers have proposed that over-activation of a class of immunosurveillance cells known as dendritic cells are causing a disruption in the autonomic nervous system.

Another theory is that chronic pain could arise from changes in glial cells that surround and support neurons. Increasingly, research is showing that pathological changes in glial cell activation, called gliosis, is implicated in chronic pain conditions – specifically that the 'crosstalk' between glial cells and neurons is part of what gives rise to chronic pain. Previous studies in animal models found that inhibiting gliosis correlated with a reduction of pain-related activity. But how we can use that information to help people who are suffering is, at this point, again – you guessed it – unclear.

What we do know is that changes in the nervous system that result in chronic pain are happening at a cellular level. The neurons' adequate stimulus – the threshold for firing – has lowered, making them more sensitive. These changes can result in nociceptive signals that should have been stopped and dealt with at the second relay station, the dorsal horn, instead getting passed all the way up to the brain. In some cases, this results in nerves in the PNS and CNS continuing to send messages about limbs that aren't even there any more – so-called 'phantom limb' pain, which affects a significant number of people with limb amputations. Chronic pain can also result from the reverse: signals from the brain that work to inhibit nociception aren't getting there – if our pain signalling is messed up, there's a good chance networks regulating our endogenous painkillers are working with incomplete or just plain wrong information.

Knowing that some kinds of pains are the result of faulty wiring might also offer up some novel treatments that aren't pharmacological. In 2018, the *British Medical Journal* presented a case study of a 28-year-old man who, after years of living with excessive facial flushing, had undergone endoscopic thoracic sympathectomy to cut the triggering nerves in his chest. The surgery worked, but after the operation he was left with a persistent, searing pain in the intercostal nerves located between his ribs. The prescribed physical therapy and movement aggravated the pain; nothing he did seemed to work and his quality of life plummeted. A triathlete and open-water swimmer before the operation, he decided to go for a swim in the sea to try to take his mind off the pain.

It was a lot colder than he expected - in fact, it was about 11 degrees. 'I initially thought, "Damn, this is so cold I'm going to die!" and I just swam for my life,' he told the report's authors, Dr Tom Mole of the University of Cambridge's Department of Psychiatry and Dr Pieter Mackeith of the University of East Anglia School of Health Sciences. But when he finally clambered out of the water, he realised he was pain-free for the first time in months. And he stayed that way. Mole and Mackeith cautioned that this was a single case study and that it's not totally certain the cold-water swim was related to the remission. But they also suggested that the shock of the cold water might have generated a burst of sympathetic nervous system activity, a power surge that reset his nociceptive thresholds.

Pain in the brain

At this point, you might be wondering what exactly happens when – *if* – the nociceptive signal reaches the brain.

Though the cultural obsession for locating where fear/love/hate/jealousy/desire to eat McDonald's fries 'lives' in the brain seems like a modern one, parsing out mental activity to different parts of the brain is not new. Phrenology was a hugely popular pseudo-scientific pursuit in the 18th and 19th centuries that mapped faculties of the mind onto bumps and shapes of the skull: 'destructiveness' is located above the left ear, 'domestic propensities' are above the right; 'desire for liquids' is somewhere near the left temple, while the 'love of children' and 'love of animals' are crowded together at the rear of the skull.

Now, however, feeling the bumps on the *outside* of a person's skull to gain insight into their character has given way to feeling around on the inside of their skull for the same information. And with advances in neuroimaging, it seemed like scientists could finally do just that, starting by mapping the folds and forms of our squashy, grey 3-pound brain to their functions.

Except it doesn't really work like that. Though there is a consistent physical anatomy to the brain, and patterns of activation that correlate to certain behaviours, the inclination to carve it up into discrete modules that *do stuff* is misguided. Rather than thinking of the brain as a collection of function-specific regions, or even as independent neural networks, recent neuroimaging science posits the brain as a dynamic and complex system of neurons and glial cells that forms one awesomely interconnected network.

Picture a typical high school canteen in a movie. Tables are populated by defined groups: the jocks, the goths, the theatre kids, the nerds, the cheerleaders, the future pop and hip-hop superstars, the loners, and so on. These groups are tied together by their shared primary interest and are densely connected, meaning they all know each other. But it's not unusual for there to be overlap between groups – the goth kids who are into theatre, the nerds into sports, the arty kids who are into politics, and those kids who know everyone and act as social connectors. Some groups stake their claim to certain spots because they need the space to do their thing, the way the gamers snag the corner with all the power sockets. Some groups care less about location and more about their proximity to other groups they hang out with the most – the cheerleaders and football players, the band kids and school choir.

We can think of our brain as organised in a similar fashion. Groups of neurons with special properties are organised into nodes; if they are densely connected, they then form hubs that constitute core systems in our brain, involved in many functions. And, like the popular, social connector kids, some hubs are more connected than others – the really connected ones are densely embedded in the brain and spinal cord, enabling better and faster communication. This type of organisation makes a lot of sense from an efficiency standpoint. Imagine someone wanted to get the word out to everyone about a party happening that weekend. They would save a lot of time and energy by recruiting the social connectors to spread the word, rather than going around from hub to hub. From this complex network perspective, then, it's not always about what part of the brain is for what, but rather what models of network activity are the best at predicting different mental events, behaviours and sensory experiences, including pain.

With that in mind, we're not likely to figure out 'where pain lives in the brain'. But we can start to glimpse a pattern of interconnected neural activation that correlates to nociception, a more accurate but less catchy description. In 2013, Tor Wager, a neuroscientist at the University of Colorado, published a paper in the *New England Journal of Medicine* describing what he called the 'neurologic signature of physical pain' – a specific and predictable pattern of brain activity across multiple neural regions that correlated with nociceptive activity.

In the initial study, he and his colleagues used fMRI to find a pattern of brain-wide activity that reliably tracked with increasing heat stimulation. The areas of the brain involved included the bilateral dorsal posterior insula, the secondary somatosensory cortex, the anterior insula, the ventrolateral and medial thalamus, the hypothalamus and the dorsal anterior cingulate cortex. Even more significant, this signature was reliably associated *only* with thermal nociceptive input: 'We identified an fMRI-based neurologic signature associated with thermal pain, discriminates physical pain from several other salient, aversive events, and is sensitive to the analgesic effects of opioids,' they wrote. 'This signature consisted of interpretable, stable patterns across regions known to show increased activity in association with experimentally induced pain, hyperalgesic or allodynic states, experimentally induced acute pain in patients, and experimentally induced tonic pain (pain caused by a stimulus of extended duration) in healthy persons.'

Since that study, Wager says they have replicated their findings in other settings, as well as teased out the pattern from where it is entangled with other emotional experiences – the pain of a burn, say, as distinct from the similar neurologic patterns the brain exhibits during the pain of social rejection. 'I'm very confident that we have a measure that responds to many types of evoked pain, heat shock, mechanical, laser, pressure, cold, visceral stimulation, rectal, esophageal, gastric, it responds to evoked painful stimulation in all of these,' he told Linda for an article for the *Boston Globe* in 2017. 'It has a sensitivity and specificity profile across multiple types [of stimulation] that's better than I would have imagined when I started to do this work.'

The findings were explosive in the world of pain research. If Wager's research was accurate, then finally there could be an objective measure, a biomarker, for the experience of pain. For individuals whose ability to communicate is somehow impaired – infants, for example, or the elderly or anyone living with cognitive impairments – his findings would be hugely important. However, not everyone agreed that this pattern of brain activity could justifiably be called nociception's neurologic signature; some leading pain researchers worried that Wager's signature was not distinct enough from other salient experiences and that all they'd managed to capture was that something sensorily significant was happening, but not specifically and *only* nociceptive pain.

Others worried about the implications of a neural signature for pain – what if someone reported being in pain but their brain seemed to disagree? Were they still in pain? This isn't just a theoretical worry: after the publication of Wager's paper, an American company called Millennium Magnetic Technologies began selling its services, positing their fMRI-based scans as a kind of lie detector for pain. As of 2016, the company had already been called in on several personal injury lawsuits to determine whether the individuals suing for damages were actually in pain. In 2017, Stephen McMahon, one of Europe's leading pain researchers and a professor of physiology at King's College London, was asked by a British high court whether this kind of scan should be admissible evidence; he said absolutely not. 'That is just pie-in-the-sky at the moment,' he told us. We reached out to Millennium Magnetic Technologies for this book, but heard nothing back after an initial response, and it seems as if they're no longer active.

Wager agrees that there is 'a lot of reason to be really cautious', echoing concerns that using brain imaging as part of diagnostics might result in patients' reports of pain being less believed, not more. 'But the question is, where do you go from there? Do you stop doing neuroscience of pain? Or do you come up with what this means and doesn't mean … what are the limits of it, what does it work for?' he continued. 'Any brain measure or biological measure, any technology can be misused … what we have to do is be smarter and more thoughtful and compassionate about how we use them.'

Wager's research underscores perhaps our most important point about nociception: it's not pain. This might come as a bit of a surprise after the last several pages. But nociception, the physiological process of stimuli-driven neuronal firing, is neither sufficient for, nor does it encompass all that is, the experience of pain. Nociception isn't even always a reliable indicator of physical injury – remember the builder we mentioned in the introduction? There was no nail puncture, no *real* damage. But his pain, at that moment, was as real as the enormous nail he'd just stepped on. The fact is, you don't even need nociception to experience pain; you certainly don't need actual tissue damage to experience it either.

What Wager has identified is a predictable, objective pattern of neural activation that occurs in relation to noxious stimuli, not a way to objectively field or assess a report of pain itself. One of Margee's colleagues, Greg Siegle, a neuroscientist at the University of Pittsburgh who uses neuroimaging to understand emotional cognition in people with mood disorders, offered a morbid yet illustrative analogy. Scientists have identified consistent patterns in the brain that occur in response to seeing a smile. But how you interpret that smile, how it makes you *feel*, depends on the context: is this person smiling while cuddling an adorable kitten or after they just drowned one? 'There's a smile network that activates both times, but your experience of that smile will be entirely different based on the context,' Siegle said. 'Pain is like that. There *is* a pain network, but there is all sorts of psychological mediation that goes on once that pain network is activated.'

So no, Wager says, this pain signature isn't measuring pain, exactly. 'I think that we're observing the processes that go into creating that experience,' he explained. 'But,' he cautioned, '[y]ou're not measuring pain, you're measuring a physiological system that people often correlate with pain.' What happens from there, whether a person perceives an experience of pain, is far more complicated, far more subjective. 'Our emotional lives are so much more complex and nuanced than what we can track in the brain,' said Wager.

This distinction between nociception and the experience of pain actually dates back to the birth of the word. British neurophysiologist Charles Sherrington's *The Integrated Action of the Nervous System*, published in 1906 and based on a series of lectures he gave at Yale University, was a terrifically influential book, outlining some fundamental ideas about the nervous system and how it works. But perhaps most significantly – at least for us – was that in it, Sherrington coined the term 'nociception' to refer to the brain's processing of injurious events, both real and potential. In one word, Sherrington gave the scientific community a way to talk about the processing of noxious stimuli as distinct from the much more complex experience of pain. 'Mind rarely, probably never, perceives any object with absolute indifference, that is, without "feeling" … affective tone is an attribute of all sensation,' he wrote.

Take it from a neuroscientist (or three): pain is emotional. In fact, we'd go so far as to say that pain *is* an emotion.

How pain is made

However, though Sherrington cautioned against depending solely on nociception to understand the experience of pain, his insights went by the wayside. Why was the role of 'affective tone', as Sherrington put it, discounted in the experience of pain? For that, we have Rene Descartes to blame.

One of the biggest victories for science and rationalism might also have kicked off more than three centuries of problematic thinking about pain. In the 1630s, Descartes, the French philosopher best known for declaring 'I think, therefore I am', posited a theory of mind that essentially cleaved the body from the soul for ever after. Descartes reasoned that things in the external world impacted the sense organs, which then registered in what he called the 'pineal gland', a sort of central processor in the human brain. The mind and the body are conjoined, he said, but ultimately separate and unequal partners in the creation of existence. In the short term, this division was a great thing for medical science – in dividing the mind from the body, Descartes made the body an area of objective study, and promoted rational observation over speculation.

However, he also laid the groundwork for the biomedical model that has largely dominated the practice of medicine since the 19th century. Humans, the model goes, are biological organisms whose constituent parts can be explained by hard sciences – chemistry, biology, physics. Disease, illness, pain are simply malfunctions of these constituent parts, and it only requires the right corrective agent to get everything back on track. The mind, meanwhile, was a separate dominion all together, requiring its own sciences – psychology, psychiatry, philosophy.

Based on the Cartesian model, medicine reduced pain to tissue damage or injury; this is why nociception has largely dominated biomedical explanations for and investigations into pain. But in practice, this also meant that if the pain reported didn't seem to have a clear aetiology or observable cause, or couldn't be neatly filed under the heading 'nociception', then either it wasn't pain or it was the product of a disordered mind. In other words, the pain wasn't 'real', and this paradigm enabled some doctors to dismiss patients' reports of pain in favour of their own expertise. The enduring legacy of Descartes' forcible divorce of the mind from the body, as we noted in the introduction, is that when someone says, 'It's all in your head', it's not meant kindly. Which is a bit of a problem. Because as it turns out, a lot of this experience that we call pain *is* in your head. That doesn't mean it's not real.

We know intuitively that our affective tone – or in today's lingo, just affect – is a reflection of how we *feel* in our bodies. Every day, how we feel is manifest in, well, how we feel; there's a reason why this one word has overlapping meanings. When we're nervous, our hearts might beat faster or we sweat; when we're sad or joyous or watching kitten videos on YouTube, we might cry. Intense emotions, such as grief, are experienced all over – they can hurt. But even describing the quotidian ways in which our emotional state has tangible physical correlates, it sounds like we're implying that the body is a thing to be acted upon by the mind. It's not. The mind *is* our body; together, they are a single complex and integrated network. Which is why it's useful and necessary to think of pain as an emotion.

We are, of course, not the first or only people to consider pain an emotion. The ancient Greeks, for example, used the word *algos* to describe pain. As in modern English, this flexible word encompassed both physical pain as well as suffering, woe, sorrow and distress. But unlike us, the Greeks made less distinction between physical pain and what we'd call emotional pain; these functioned in largely the same way. So while they could talk about pain as resultant from, say, an arrow to the chest, this was in a similar category as the pain that came from finding out it was your brother who fired the arrow.

What the Greeks didn't know then and what we do know now is that the areas of the brain involved in processing nociceptive pain are also implicated in processing social pain. In fact, over-the-counter drugs used to inhibit nociception also blunt our emotional responses. A widely reported 2015 Ohio State University study found that people who took medication containing acetaminophen reacted less strongly to emotionally arousing images than people who took a placebo. That pain we'd put in the emotional category and the pain we'd put in the physical category share some neural circuitry, and that the same drugs can be used to impact both is evidence that, at the very least, pain is the product of multiple integrated processes.

If we think of the experience of pain *also* as an emotion, we can understand pain much more fully. Rob Boddice, author of *Pain: A Very Short Introduction*, offers this example of the way pain is an emotion: 'If I fall and break my arm while out walking in a forest, I might complain that it hurts, but whatever the nature of my physical pain, that hurt will always also express fear, alarm, and an agitated state of planning for action.' Considering pain as an emotion acknowledges its dimensions – its subjectivity, its profundity and, crucially, its malleability – in a way that restricting it to solely nociception or another rigid definition can't. It also opens up far wider avenues for managing and treating it.

But what, exactly, is an emotion? Like pain, like obscenity, like pretty much any other subjective concept, we probably think we know an emotion when we see one. In fact, the 2015 Pixar film *Inside Out* is built on the notion that not only do we all know what emotions are, but that these are also universal discrete, observable, objective states. In the film, the emotions joy, sadness, disgust, anger and fear are crafted into endearing characters with their own motivations and emotions (how's that for a mindf*ck?). It's a great movie but, as Lisa Feldman Barrett wrote for Boston Public Radio station WBUR after the film came out, it gets something deeply wrong about emotion – and so do a lot of scientists.

Feldman Barrett is a professor of cognitive neuroscience at Northeastern University and the author of *How Emotions Are Made*, a insightful book that we think nails emotion. The problem with *Inside Out* is that it hinges on the idea that emotions are singular, discrete, innate states with rigid boundaries (and personalities). The problem with some scientists, wrote Feldman Barrett, is that they 'have taken this model seriously for a century and actually search for these characters in the brain.' Not as cartoons but as 'blobs of brain circuitry', consistently identifiable from person to person. 'This blob over here is your "fear circuit," they say, or this other blob "computes anger." And every time you experience an emotion, your corresponding blob of neurons supposedly leaps into action, triggering your face and body to respond in a consistent way,' she wrote. This narrative is fundamentally flawed: 'Today's neuroscientists finally have the technology to peer into a living brain without harming its owner, and it's clear that the brain doesn't operate even remotely in this cartoonish fashion. We might perceive Joy, Fear, and Anger as separate entities – even gloriously rendered in 32-bit color – but the evidence from neuroscience is overwhelmingly against it.'

Again, finding where something 'lives in the brain' is reductive and unrealistic. Feldman Barrett's lab has found, through meta-analyses of brain-imaging studies, that though there are brain circuits for behaviours such as fight-or-flight reactions, 'no brain region is the home for any single emotion.' Complex mental states such as fear, anger or joy are, well, complex and 'don't have consistent responses in the body either.' The success of *Inside Out* was in part due to the satisfaction that many of us felt watching our messy, multi-layered emotional experiences transformed into colourful cartoons who could interact, communicate and problem-solve independently. But distilling our emotions to discrete blobs of brain activity is dangerously misguided. 'The blob-ology of emotion would be cute if it weren't also so serious,' wrote Feldman Barrett. 'The U.S. Transportation Security Administration spent almost a billion dollars training its agents to recognise terrorists, on the assumption that their facial and bodily movements could reveal emotion. (It failed.) Our legal system at its core treats emotion and reason like two battling characters in the brain. (They're not.) Medical researchers investigate the relationship between heart attacks and anger, as if anger has one consistent state in the body. (It doesn't.) So many critical parts of our lives rest on the invalid assumption that emotions can be located distinctly in the brain.'

In her book, Feldman Barrett offers an alternative theory, one that understands emotion as a constructed experience built from affect, sensory information, cognitive processing, our memories of previous experiences, context and expectations. We like this theory, largely because we believe it tallies with lived experience. What we call 'emotion' is an organising concept under which is an endless vocabulary of words to describe feelings – pain, joyfulness, fear, relief, disgruntlement, anger, bemusement, hopefulness, hanger, wonder. We learn or create these emotional concepts to further organise similar – but never exactly the same – combinations of sensory information. This means each instance of pain is constructed anew in the moment. For example, even though monthly menstrual cramps or throbbing headaches may feel the same, each one is a brand-new instance, similar in ways that are meaningful but not the same. As the saying goes, we may swim in the same river but we'll never swim in the same water. This is important because when we understand that emotions and pain result from an active process of construction, we can see the opportunities to change how an experience is built.

Emotions may be constructed from what we learn, but recent studies of infant brain activity suggest we're all born with the ability to experience *affect*, which Feldman Barrett defines as your 'simplest feeling that continually fluctuates between pleasant and unpleasant, and between calm and jittery.' Another way of thinking of affect is as a mental representation of what's going on in our body. Constructing an instance of an emotion starts with affect.

We are always feeling some kind of way – it's a condition of being conscious – and our feelings are in perpetual flux; the only time we are in a static state is when we are dead. Though it's always in a state of change, we can plot affect on an x–y axis: the x axis describes what affective researchers call 'valence', the range between 'ugh' to 'ah', or pleasant to unpleasant; the y axis is arousal, a line between revved up and ready to pass out. The locations within this axis can be described with an emotion concept, such as excited, surprised, scared, angry, content; you could even assign each of these locations one of those handy modern pictographs, emojis, just for visual ease and kicks.

Every moment we are alive, we are moving within this circumplex, propelled not just by the things that happen to us or what we happen to be thinking about, but also by the underlying systems that support basic functioning. Think of all that is happening in our body at any moment: our exteroceptive networks (that is, nociception, proprioception, somatosensory perceptions) and our interoceptive networks (specifically our autonomic visceral and vascular systems, neuroendocrine and neuroimmune systems) are constantly processing information, balancing resources and figuring out what we need and how to get it in the most efficient way possible. This ongoing process is called allostasis, and according to Feldman Barrett and her colleague, W. Kyle Simmons, 'Allostasis is not a condition or state of the body – it is the process by which the brain efficiently maintains energy regulation in the body.'

Regulating the body – ensuring the right systems have the right resources – is central to functioning, so much so that resource efficiency is kind of our organising principle. But, as sophisticated and sufficient as they are, these systems that keep us alive cannot generate resources – fuel – by osmosis. Something has to motivate us to put food in our mouths, seek warmth when we are cold, protect ourselves when we're injured. In other words, it's important to feel bad when our body needs something. It makes sense and seems almost too obvious to mention, but consider how far we, or any other species, would make it if we were able to perceive a gurgling stomach but didn't really care about it?

Or consider this, since this is a book about pain: what if we were able to perceive nociception, but had no interest in doing anything about it? Pain asymbolia is precisely that and it can be just as dangerous as insensitivity to pain. People with pain asymbolia, a condition that can arise from trauma to the brain, know that a sensation is painful, it just doesn't *mean* anything to them. One of the earliest mentions of pain asymbolia in scientific literature concerned an Austrian woman who 'pushed everything that came into her hand against her eyes, heedless of the pain she thus inflicted upon herself.' This was 1927, in Vienna; it's unclear from the case report why the woman was under their care, but Dr Paul Schilder and Dr Erwin Stengel, the two attending physicians, were sufficiently intrigued by her behaviour to observe her closely. They found that she 'did not react to pain or only in an incomplete and local way', and that, for the most part, 'There was no real defence action … She never became angry at the examiner and even seemed to derive some pleasure from the pain. Sometimes she took a needle and stuck herself deeply with it.'* Other case studies of individuals with pain asymbolia also noted that patients were not only unconcerned about painful stimuli but also exhibited no reaction

* The two were so intrigued by her case that after her death – notably, they did not mention how the woman died – they performed a detailed autopsy of her brain. They found lesions peppering her left brain hemisphere, in the temporal and parietal lobes. But most significantly, they thought, she had large lesions in her 'gyrus supermarginalus'. The supermarginal gyrus is a portion of the parietal lobe and part of the somatosensory association cortex. It's also implicated in things like language facility, empathy and facial recognition.

to threatening gestures and words.* Pain asymbolia is extreme and unusual, but it demonstrates a primary function of sensory perception – to influence affect with the goal of motivating behaviour. After all, in order to survive you've first got to care enough.

Having a sense of feeling good or bad is necessary to our survival, so it makes sense that we all are born with the ability. However, the meanings we ascribe to the rich diversity of our feelings are something we've learned. By the time we're adults, we've learned a lot about the meaning of our physiological sensations, the differences between a hangover and the flu, between feeling hungry and feeling tired (or, god forbid, feeling both at the same time). But when we're born, our physiological sensations, the things happening inside our body, have no inherent psychological meaning beyond how it changes our affect; as soon as we're launched into this world, however, we're learning how to assign them meaning, even before we have the words to express them. A baby doesn't cry because she has an understanding of what it means to be 'upset' but because her quickly developing body needs resources and she hasn't been fed for hours. Her affect is in the gutter and something needs to happen to make things better. As soon as the missing ingredient is found and delivered, things start to feel all right again and the urge to cry diminishes (unless it doesn't, which happens a disturbing amount with babies because they are terrible communicators). With each of these experiences, the baby begins to associate physical sensations with specific needs, how those needs are met, and the words, people and places around them when it happens. The baby starts to make meaning.

This work is happening from the very beginning. A 2007 study published in *Developmental Psychology* found that four-month-old infants understood the difference between pleasant and unpleasant affect in other people. Considering four-month-old infants have yet to learn the distinction of self and other, it's safe to assume their behaviour suggests they also understand pleasant and unpleasant

* Pain asymbolia can be induced and, for a brief and terrible period in medical and psychiatric history, routinely was: lobotomy, basically poking an ice pick into the back of someone's eye and swizzling it around in their frontal lobe. More on that in the next chapter.

within themselves. But they couldn't do it without a little extra context: infants couldn't tell the difference when they were shown only a person's face, they needed vocalisations like 'Oh, hi, baby, look at you! You're such a beautiful baby!' As a baby is held and soothed, with food, cuddles or rocking, the baby – and usually the caretaker – feels better.* This simple basic associative learning is what we scaffold, or build language and emotional regulation on, replacing cries with self-soothing skills and words. These patterns are organised into categories, then mapped onto mental concepts, concepts from the language we learned, a language created in and by the culture in which we are socialised.

The fact that a facial expression alone was not enough information for infants in the 2007 study to tell the difference between good and bad highlights how learning, especially early learning, is a multi-sensory process. Feldman Barrett writes, 'An infant's interactions with the wider world around her – including other people and the words they speak – function as a set of instructions for her brain to wire itself to the physical and social conditions of her environment, a process that takes more than two decades to complete.' Our early years involve mapping out what constellations of physical sensations and environmental contexts result in what kinds of desirable or undesirable outcomes. This process is called statistical learning, and our brain is really good at it.

Feldman Barrett describes statistical learning as 'an inborn ability of the brain to learn patterns by observation, computing probabilities of what is similar and what is not'; statistical learning is purposeful in that we are creating patterns that are meaningful because they help us get what we need. But we need some way to organise and hang on to the patterns that we learn, otherwise what good is learning them? This is where language comes in. Like magic, a mental concept defined by a word – or even a series of gestures and movements, such as sign language or body language – has the power to tie all of these

* Research shows that children who exhibited physiologic synchronicity (similar heart rates, skin conductance) with their parent displayed better self-regulation, adapted behaviours more easily and showed a greater capacity for empathy in adolescence.

context clues together. Evidence suggests that even before we know what language is, babies learn that specific vocalisations are part of specific patterns of sensations and contexts. Over time, fewer and fewer context clues are needed until just one word is enough to capture and communicate a boatload of information. And it is far less taxing on our body to be able to use a single word or gesture as a shortcut to get what we need instead of screaming our heads off (at least for most of us).*

As adults, all we need is a picture or just the tone of someone's voice to make a pretty accurate guess at whether they're feeling good or bad. Being able to quickly and mostly accurately infer what others are feeling is a kind of emotional cultural synchronicity. If this wasn't the case, and we needed a combination of visual, auditory and tactile clues or even the person explicitly stating how they feel in order to infer their feelings, everyday interactions and communication would be exceedingly difficult and resource-intensive. To get an idea of how much meaning we internalise through socialisation, try travelling to a different culture where you don't speak the language.

We, as individuals, do not create meaning independently. Meaning is socially constructed from the continuing co-creation of knowledge that occurs between people and groups. Through this process norms, social values and beliefs emerge and are then taught to us in schools and in playgrounds and internalised through experience. Social and cultural norms vary by time and place and some of those differences are easily recognisable – consider the variety of socially appropriate greetings the world over, from kissing on the cheek to shaking a hand to a bow. But there are many often imperceptible and automatic norms of thinking, feeling and acting that we've internalised simply through the process of growing up in a culture.

How we express pain and are therefore primed to respond to other people's pain is, in part, a product of this co-creation. Amanda

---

* Another very good example of statistical learning is the human tendency to see faces in grilled cheese sandwiches, parking meters and clouds: 'pareidolia' describes the inclination to ascribe 'faceness' to things that are not faces, on the evolutionary wisdom that two eyes, a nose and a mouth are typically a face. It's also why emojis are recognisable as faces.

Williams, then a lecturer and a consultant psychologist at the INPUT Pain Management Unit at St Thomas' Hospital, suggested that human expression of pain arose from evolutionary propensities in a 2002 paper for *Behavioral and Brain Sciences*. Certain pain behaviours conferred evolutionary advantage on individuals who used them: acute pain is often associated with reflexive movements, vocalisations ('Ouch!'), grimaces and facial expressions that elicit sympathy and possibly aid from the observer, as well as alert others to potential threats.

'Of necessity, vigilance to observed pain cues in others coevolved. The individual expressing pain would derive benefit if expression of pain were reliably followed by actions by observers that promoted recovery and survival; protection from danger; and aid in obtaining basic requirements,' explained Williams. 'If the person in pain might survive rather than not, and the cost to helpers is low, selection advantage follows as with other help and exchange of information – the currency of kin or reciprocal altruism.' In order for pain expression to be effective, it has to be understood by others around us. These expressions are not 'hardwired' and there isn't a set of universal facial expressions that 'mean' pain, they are the products of reinforced cultural norms passed generation to generation.

We create emotional concepts and learn the language of our culture so that we can communicate, learn, get our needs met and survive. But because we're creating and applying these shared labels to our subjective experience of how we feel at any given moment, we will never know if what pain feels like for you is the same as what it feels like for your best friend.

Think about the concept of grief. Like many emotional concepts, we often talk about grief by invoking other emotional concepts – in English dictionaries, the word grief is flexible enough to accommodate both 'deep sorrow' and 'minor annoyance'. Your experience of grief, how it feels and how you know you're feeling the deep sorrow version of grief or the annoyed version is known only to you. A sense of grief for someone who lost a beloved parent at a young age might be very different from someone who has yet to see any loved ones pass. Neither person's experience of grief is more right, but if they were raised in the same culture around the same time, it's likely that they share enough common understanding to have a conversation and to make a pretty good guess about what the other is experiencing. But the fact remains that each of our understandings of the word will never be completely known to anyone else – that is the problem and the intrigue of other minds, and the ineffable nature of our inner experiences.

The variability of grief highlights another important point: mental concepts are not exclusive or rigidly defined, and there can be a lot of overlap between affective properties. An upset stomach and a waterfall of tears can characterise the experience of grief, but they can also characterise an experience of anxiety or even strong joy. Or another example: a racing heart, sweaty palms and rapid breathing can be a part of feeling fear but can also be a part of feeling sexy, excited, nervous, in pain or all of those at once.

The diversity of emotions is perhaps best illustrated by the fact that every culture has words describing emotional experiences that cannot be translated exactly, or even words to describe emotional experiences that may not exist in another culture. You're probably familiar with *schadenfreude*, a German word meaning 'pleasure from another's misfortune', or *hygge*, the Danish and Norwegian concept of convivial cosiness that launched a thousand Pinterest boards, but there are so many others. A favourite of ours is the Tagalog word *gigil*, which describes a feeling of such intensity that it challenges your self-control. It's often used in response to seeing something considered adorable – that feeling of 'Oh my gosh, you are so cute, I just want to bite you!' – but it also can be used to describe trembling with intense anger or frustration. When the Finnish talk about their national character, they use the word *sisu*, a combination of resilience, stoicism and persistence, a gritty determination that underscores courage. In Iran, interpersonal interactions are often rules by the concept of *Ta'arof*, a Persian word describing the art of etiquette that means sometimes people refuse when they want to accept and offer what they can't give. English doesn't have a single word that translates these concepts directly, and no matter how many other words we use to describe them, we will never know them exactly as those who grew up with them or have internalised them through assimilation. Even when we share a common language, acculturation lends those words different weights or meaning – consider how the British use the word 'quite' versus how Americans do. If something is 'quite good' in America, it's very good or even great, but in Britain 'quite good' isn't really good at all.* This is because we learn these emotional words through interaction, through instances where it is used by others in specific contexts.

Understanding emotions – pain included – as constructions built on affect, shaped by our experiences and given meaning with the help of those around us, the culture in which we are raised and our genetic inheritance brings both great opportunity and consequence. First, the opportunity: that emotions are not discrete states but constructed in the moment means they can be *re*constructed. The way we experience pain today does not mean that it will, or has to be, the way we experience it tomorrow. We can't go back in time and change previous experiences, or our genes, but there is so much we *can* change. The meaning we give to the way we *feel* is more in our control than many think, a truth that will be revealed in the pages and stories ahead.

Now to the consequences: we cannot know pain (or joy or fear) in others exactly as they know it in themselves, just as they cannot know it in us. We can make really good guesses, we can empathise – using our own pain to try to better understand theirs, we can glean, infer, predict, essentially every word other than 'know' – and most of the time we'll get pretty close because there's a strong synchronicity between our own definitions and those defined by the culture in which we live. But not always. The consequences go beyond our ability to talk and connect with our friends. Applying our understanding of pain to another's experience might – and often does – lead to misunderstandings, misdiagnosis and mistreatment. In those moments, the struggle to have one's pain recognised as real, to be heard and believed, is a source of pain in and of itself.

* Linda here: both my husband and I were born and raised in America, but our children are British. This is never more apparent than when it comes to the word 'quite'.

The 'definition' of pain

Definitions matter. The definitions of pain that are accepted as a social reality – that are codified by power holders through official definitions, diagnoses, reimbursement schedules, disability payments, court rulings and treatment protocols – matter. But too often, the definitions used by these stakeholders are the ones that suit them best, and not necessarily the individual in pain.

In 1979, the International Association for the Study of Pain (IASP), responding to what it felt was a need to address the problematic treatment of pain, issued a definition of pain that would become the most widely used by medical practitioners and researchers. Pain was, the organisation said, 'an unpleasant sensory and emotional experience associated with actual or potential tissue damage, or described in terms of such damage.' Sounds simple, right? Not exactly – the definition came with a 250-word footnote, dense with caveats and qualifiers. At the heart of the note was this sentence: 'If they regard their experience as pain, and if they report it in the same ways as pain caused by tissue damage, it should be accepted as pain.' The assertion that the individual had the right to declare their own pain was a relief for the millions who might have felt invisible; coupled with the acknowledgement that pain is also an emotional experience, the definition went a long way to validate pains that were often dismissed by the biomedical establishment.

However, there were problems with the definition right from the start. Though it acknowledged the role of experience in pain, it failed to include social or cognitive factors in that experience. Though the definition noted that the 'inability to communicate verbally does not negate the possibility that an individual is experiencing pain', critics claimed that this didn't do enough to address pain in disempowered populations, such as infants, the elderly, and individuals whose ability to communicate is impaired. And though the definition recognised that pain had emotional dimensions, one phrase in particular seemed to affirm the Cartesian divide and undermine an integrative approach to pain care: 'Many people report pain in the absence of tissue damage or any likely pathophysiological cause; usually this happens for psychological reasons.' No matter the intent, the phrase 'psychological reasons' can sound like 'It's all in your head.' Essentially, it sounds like 'Get over it,' or, 'Your pain isn't *real*.'

The definition reflected progress, but it still wasn't *it*. In the spring of 2018, the IASP decided that it was time to re-examine their definition of pain. It took the committee – a multinational, 14-member body representing practicing clinicians and pain scientists – two years to come up with a revamped definition, published in September 2020.

According to the new definition, pain is 'an unpleasant sensory and emotional experience associated with, or resembling that associated with, actual or potential tissue damage.' So far, not terribly different from 1979. But, just like in 1979, the real meat was in the note that accompanied it:

- Pain is always a personal experience that is influenced to varying degrees by biological, psychological, and social factors.
- Pain and nociception are different phenomena. Pain cannot be inferred solely from activity in sensory neurons.
- Through their life experiences, individuals learn the concept of pain.
- A person's report of an experience as pain should be respected.
- Although pain usually serves an adaptive role, it may have adverse effects on function and social and psychological well-being.
- Verbal description is only one of several behaviors to express pain; inability to communicate does not negate the possibility that a human or a nonhuman animal experiences pain.

In a lot of ways, the new definition represents how far we've come in recognising pain as complex, arising from a host of factors, and ultimately defined by the individual. But reaching this definition also represents both how valuable defining pain is and just how hard it is to do.

In the two years following the formation of the task force, the committee took apart and examined the original definition, establishing the three core principles that the new definition would rest on: that it should be valid for all chronic and acute pain conditions, should be applicable to human and nonhuman animals, and should be defined from the perspective of the sufferer rather than the observer. What had been explicit in the 1979 definition was now central to the new definition: because the issue of who has the authority to decide who is in pain – medical doctors, individual sufferers, politicians, government, the law, insurance and pharmaceutical companies, scientists – is 'contentious', the report notes, 'primary focus of the definition should be that of the individual experiencing pain.'

The task force recruited experts from across academic disciplines, including philosophers and linguists who could advise on the translatability, potential connotations, and nuanced meanings of the definition's language. And crucially, they opened up comments on the proposed definition to the public. A total of 808 respondents, 58 per cent of whom were professionals and about 42 per cent who identified as patients or care providers, from 46 countries completed the survey. Reflecting either a case of true compromise where, as the saying goes, neither side walks away happy, or the reality that pain is ultimately beyond an explicit definition, the survey revealed that 41.7 per cent were satisfied or very satisfied with the proposed definition – only slightly more than the 41.5 per cent who said they were dissatisfied or very dissatisfied.

The definition ultimately adopted by the task force and then the rest of the IASP *is* a compromise. There are concerns about the centrality of words 'tissue damage' in the definition, concerns that neuropathic pain is inadequately acknowledged, that pains that cannot be traced back to an injury are still regarded as less legitimate. The new definition is also marked by what else it doesn't include. Though the latest version has thankfully, finally thrown out the stigmatising phrase 'psychological reasons' and further points out that pain cannot be inferred 'solely from activity in sensory neurons', it also lost a phrase between the 2019 draft and 2020 final version that would have been incredibly useful to people in pain. In the draft version of the definition, the fourth note read, 'A person's report of an experience as pain should be accepted as such and respected.' The inclusion of 'should be accepted as such' would have further added to the patient's authority, acting as a bulwark against, for example, a GP respecting a patient's pain but not accepting it.

According to the report the task force produced, members also debated including what they termed 'social injury', which would have included the pain of psychological trauma or abuse as 'clinically important forms of chronic pain'. Why social injury didn't make it into the final definition isn't clear. However its relevance in the experience of pain is, we'd say, inarguable, not least because of the impact of trauma and other adverse experiences on how pain is felt. (Spoiler alert: that's Chapter 4!)

Only time will tell how this definition will be accepted, adapted, or ignored by pain stakeholders. And only time will tell whether the next iteration of the definition will resolve the already evident issues – because there *will* be a next iteration. In the conclusion of the task force's report, the authors describe the definition and its note as 'a living document that is updated in concert with future progress in the field.' The acknowledgement that there must be another definition – and another, and another – is part of the more ineffable gains this revision represents. Those trying to direct the conversation about pain are finally sharing the power of the definition with the very individuals it impacts, the people living with pain.

Pain is hard to pin down. We know this, the IASP task force definitely knows this. But, like we said, definitions matter. As we've pointed out, we're still struggling to get out from under a strict biomedical model that left us with an overly narrow definition that excluded millions of people in pain. How our definitions of pain sync or don't with societal definitions carries serious, life-threatening consequences when applied to our justice system, to how we accord value to individuals in society, and the meeting of pain in treatment rooms and beyond.

Defining pain *is* hard, but that doesn't mean we shouldn't try, and that we shouldn't try to find a definition that is as dynamic, holistic, and inclusive as possible. Because if we don't, others (whose priorities might not be with the sufferer) will. As we'll see in the next chapter, who has the authority to say what is pain and what's not is in a position of real power, one that decides how pain is recognised and how it is treated. If it is treated at all.

CHAPTER TWO

Who is in Charge of Pain?

In the 1940s, Argentina's politics were dominated by Colonel Juan Perón and his wife, the charismatic actress Eva Perón. Juan became President of Argentina in 1946, while Eva, his most ardent supporter and propagandist, soon became a kind of living saint, beloved by workers and the poor and known affectionately as 'Evita'. By August 1951, Evita's supporters were pressing her to declare her candidacy for the vice presidency. Evita, though she'd been amassing political power for precisely that position, demurred. She needed more time, she said.

The truth was, time was the last thing Eva Perón had – she was dying and only a handful of people knew it. Throughout 1951, her health had been declining at a terrifying rate: she suffered fainting spells and intense abdominal pain and, in September 1951, she was diagnosed with advanced cervical cancer. Her doctors and her husband kept it from her. Soon after, she underwent a radical hysterectomy but the cancer had metastasised. On 4 June 1952 she made her last public appearance, wearing a large fur coat over a wire and plaster frame that held her upright. On 26 July she died. She was 33.

After her death, however, rumours swirled that Argentina's beloved First Lady had undergone a lobotomy months before she died. In 2011, a Yale neurosurgeon confirmed, based on contemporary reports and physical evidence, that this was true.

Lobotomies are now largely regarded as barbaric, an extreme example of medical hubris married to patriarchy. The procedure involves the severing of neural connections – white matter tracts – between the prefrontal cortex, the area of the brain charged with executive function, and the deeper parts of the brain, including the thalamic and hypothalamic areas, which loosely deal with emotion. This was typically done by drilling into the skull and using a needle to damage parts of the brain or, in the case of the transorbital lobotomy, introducing a surgical instrument very like an ice pick above the patient's eye, through the back of the cavity and into the brain.

The result is a marked reduction in manic behaviours but can also be a marked reduction in pretty much all other behaviours as well – people who underwent lobotomies were sometimes described as apathetic and childlike.

'There were some very unpleasant results, very tragic results and some excellent results and a lot in between,' Dr Elliot Valenstein, author of *Great and Desperate Cures*, a book about the history of lobotomies, told National Public Radio in 2005. But in the absence of other ways to help people with very serious mental illness, psychosurgery genuinely seemed like a miracle. It also made people whose behaviour was in any way outside of social norms easier to control, and demand for the procedure grew in the 1940s and 1950s. As much as 80 per cent of the people who underwent the procedure were women; their consent wasn't always or even frequently sought. But what made Eva Perón's lobotomy different was that supposedly the surgery was to treat her pain.

Evita wasn't the only person to undergo a lobotomy to manage pain, and it wasn't always a last-ditch effort. In a 1947 paper, Dr Walter Freeman, the doctor who championed the procedure in America, and Dr James Watts described pain as both a sensation and an emotion. As a sensation, pain has a 'threshold, a quality, and a localisation', and is transmitted through the body via the same pathways to the central nervous system as temperature sensitivity, they wrote. On the other hand, however, 'pain is an emotion, vaguely to acutely uncomfortable, like fear, disgust and sadness or longing.' Lobotomy for pain neatly straddled both definitions, a sufficiently 'medical' treatment that targeted emotional disturbances. And it worked, they said.

Freeman and Watts offered several case studies. There was the middle-aged woman suffering from 'an involutional depression with considerable agitation and a superimposed barbiturate intoxication', who 'tossed about in bed complaining bitterly of pain about the anus.' Though an examination revealed that she had haemorrhoids, she was not currently suffering from inflammation. Freeman and Watts decided her pain was the result of her mental state; the patient underwent a prefrontal lobotomy 'with spectacular results'. 'Convalescence was rapid and before the end of the month she was resuming her household duties.' She never mentioned haemorrhoids again.

Then there was the woman whose back pain had left her bedridden and emaciated, weighing just 79 pounds, and who (unsurprisingly) suffered from light-headedness, 'roaring in her ears' and a tightness in her throat; she'd also previously lost her gall bladder, appendix and uterus. After her lobotomy, however, she still had all the pains – but had stopped complaining about them: 'I don't bellyache any more,' she told them. 'It don't get you nowhere.' Or the morphine-addicted taxi driver whose tabes dorsalis, a degenerative nerve condition typically caused by neurosyphilis, left him plagued by 'lightning pains'. After his lobotomy, he got off the morphine, gained 50 pounds, and 'laugh[ed] off' the recurring pain as 'twinges'. Freeman and Watts concluded that: 'Prefrontal lobotomy has a beneficent action upon pain whether it is primarily mental or primarily physical. It does not interfere with the perception of pain, but rather with the evaluation of pain. It does not relieve the pain, but rather the disabling reaction to pain, the fear of pain.' In other words, patients were still in pain – they just didn't care any more.

Removing the fear of pain is, on the face of it, not such a bad idea, especially if the pain signal is no longer giving us useful information. The problem was that lobotomy was basically just sticking a cocktail stirrer in someone's eye, swizzling it around and hoping for the best. Lobotomies for pain also demonstrate that how we treat pain is shaped by context, by what we value as a culture and society. Freeman and Watts' tone suggests they were mostly concerned with slotting the individual back into a functional role in society. In Evita's case, her lobotomy appears to have been an attempt to control not her pain but her increasingly incendiary and politically damaging behaviour. Lobotomies for intractable pain could be considered mercy, but they could also be a despicable symptom of a culture that cared more about social harmony and control, and about how much pain could disrupt the social order, than about the individual sufferer.

Whoever defines pain and its treatment, in all senses of the word, is in a position of enormous power. For much of the 20th century, this authority to define and determine the value of pain – often in literal, economic terms – has been in the hands of people other than those who are suffering it. Medical professionals, governments, health insurance companies, pharmaceutical companies, these are the institutions that have been in charge of acknowledging and managing pain. And this concentration of power has had, in some cases, damaging consequences, especially for anyone whose experience of pain didn't fit the dominant idea of what pain was.

Consider this: up until the mid-1980s, newborn babies routinely underwent major surgery without anaesthesia. Even as life-saving surgeries became more invasive, the majority of newborns were given nothing more than a muscle relaxant to keep them from thrashing around during the operation; as the *New York Times* reported, 77 per cent of all the newborns worldwide who underwent surgery between 1954 and 1983 to repair a serious blood vessel defect 'received only muscle relaxants or relaxants plus intermittent nitrous oxide.' It wasn't until 1987 that the American Academy of Pediatrics formally declared it unethical to operate on newborns without anaesthetics. Though physicians had legitimate concerns that anaesthesia itself could harm or kill the baby, the primary reason this had happened – and happened for so long – was that the medical establishment had convinced itself babies couldn't feel pain. Studies from the 1940s supposedly confirmed that infants hadn't yet developed the neurological capability to perceive pain because they didn't seem to react to pinprick tests. Because newborns' outward expression of nociceptive events didn't match what they thought pain expression looked like, scientists, doctors and clinicians all decided this meant they weren't in pain at all. Subsequent studies demonstrated that infant responses to noxious stimuli are as well developed as older children's, but these studies remained cloistered in medical journals, and few surgeons and anaesthetists knew about them. Challenges to accepted wisdom went ignored, and, as the *Times* reported, 'Only after parents and other laymen raised a cry about needless suffering, and some filed lawsuits, was there enough pressure to change.'

Not only is the diagnosis and treatment of pain often complicated by bad or incomplete science and by inaccurate ideas about what being in pain should look like, it is also distorted by biases. For example, why are men more likely to be diagnosed with heart disease but women are more likely to die from it? Because established medical wisdom, for decades, misdiagnosed and disregarded women's symptoms: where men are more likely to experience chest pains during a heart attack, women are more likely to feel pain in their neck, jaw, shoulders and arms, and to experience nausea and vomiting. These pains – because they don't look like what physicians have been taught heart disease looks like – are often dismissed as indigestion or even anxiety.

This is not all doctors' fault, by any means, but it's a giant, glaring flaw in the dominant biomedical system. Lobotomies for pain are the kind of thing that happens when ownership of pain is concentrated in a few, and when those few are willing to brush aside the individual sufferer's autonomy in favour of their own opinion. Infants undergoing open-heart surgery without anaesthesia is what happens when we have an overly narrow definition of what pain looks like or how it is measured. And preventable deaths from heart disease is what happens when an inaccurate picture of pain becomes hegemony, baked in to how medicine is done.

So how did we get here?

How did we make doctors?

We tend to think of history, particularly scientific history, as a kind of steady, uphill climb of progress. But this isn't exactly the case. Though 'reason' has consistently won ground from 'superstition', this is less a series of battles between entrenched enemies and more a never ending succession of small skirmishes, shifting alliances and unclear outcomes. What this means is that the landscape of pain is crowded with flags – religion, science, the state, to name a few – and has been from the beginning of recorded history.

In super-broad strokes, when confronted with all the weird stuff bodies do, many cultures naturally turned to supernatural explanations. The Ebers Papyrus is one of the oldest medical treatises in the world, dated to around 1550 BC and containing the wisdom of ancient Egyptian physicians. Burns, for example, could be treated with the timely application of a frog in warm oil* or with an invocation

* Exactly how is unclear. Is the frog fried in oil and served warm? Is the sufferer meant to give a live frog a warm oil massage? Or dip a frog, alive or dead, in warm oil and then apply the even-slippier-than-usual amphibian to the affected area? And how would any of this be conveyed in hieroglyph?

of the god Horus to relieve the heat – 'O son, Horus! There is Fire in the Land! Water is not there and thou are not there! Bring water over the River-bank to quench the fire!' – to be spoken over the 'milk of a woman who has borne a son'. In ancient Egypt, medicine mingled explicitly with magic; there was no conflict between 'scientific', causal explanations for illness and injury and supernatural explanations.

But from about the fourth century BC, Hippocrates' theory of the four humours became the dominant explanation in the West for human health, while in other cultures, similar theories about internal balance also grew in influence. Hippocrates, who believed that disease, lifestyle and environment were responsible for ill health, sought to divorce mysticism from medicine; in his view, there were four bodily fluids that could have negative effects on the individual when out of balance. 'The body of man has in itself blood, phlegm, yellow bile and black bile; these make up the nature of his body and through these he feels pain or enjoys health,' *On the Nature of Man*, a Hippocratic text, explains. Mental illness and physical illness, pain, were seen as largely the same kind of thing, an imbalance somewhere, and diagnosis of any illness took into account the entire human. This idea was later expanded by the second-century physician Galen, whose prodigious written output alone meant these theories would dominate practice for about 2,000 years. Treatments designed to put the humours back into balance ranged from relatively benign, such as diet or lifestyle changes, to much more aggressive, including induced vomiting, laxatives and bloodletting.*

This, of course, didn't mean spirituality and religion no longer had a say in how bodies were healed – it just meant there was another flag in the field, planted by science. Religion continued to exercise authority over human experience; for centuries, Christians located pain and suffering solidly in the context of sin and guilt: Jesus, after all, healed not only spiritual affliction but physical as well. This

* Though this theory of health was generally disproven by the emergence of germ theory in the 19th century, its attendant practices persisted even into the 20th. No, really – bloodletting was still a recommended practice in at least one medical textbook in 1923.

implied pain had a moral value, that suffering had meaning, even if it wasn't entirely perceivable to the sufferer. As Joanna Bourke writes in *The Story of Pain*, 'In Anglo-American societies, religious dogma and practices have provided the most robust materials from which the meaning of bodily pain has been constructed ... [T]he most pervasive theological presences have been Catholic and Protestant versions of Christianity. Their engagement with bodily pain has relentlessly insisted that pain has a divine purpose. Deciphering that purpose has not been easy.' More than that, it complicated treatment: if pain had moral value, there was less of an imperative to try to relieve it. Through the middle of the 19th century, pain retained value as a moral corrective and, as Bourke argues, a social corrective, even in the abstract – pain could serve as a reminder of God's authority, as well as a reminder of an individual's place within the social hierarchy, in the great chain of being.

As Descartes' theory of mind drove a wedge between the body and the soul, religion's authority over the human body's functions gave ground to science's. (Its influence over behaviour – how bodies are used – however, remains strong.) As medicine inched towards science and causal explanations for disease, illness and pain were increasingly mainstreamed, care for the body in pain was largely, but not completely, given over to the medical doctor.

From very early on, however, there were other players too, not the least of which was the state. Care for the body – human or otherwise – has nearly always been regulated in some fashion, with varying degrees of stringency and success. The Code of Hammurabi from 1740 BC – some 282 Babylonian laws inscribed on a massive stone stele – included regulations regarding physicians, surgeons, midwives, wet nurses and veterinarians; the slab detailed their responsibilities, how much they could charge and what happened if they messed up ('Physician, amputate thyself!'). This instinct to regulate, define and award bodily authority also persists throughout recorded history.

Fast-forward a few millennia, and the interests of the state aligned with the emergence of a new authoritative bloc, itself drawing on the formalisation of scientific practice: education. From around 1200 AD, universities began popping up across Europe – Paris, Bologna,

Oxford – and many offered education in medical practice. Exactly what that education entailed varied, however, the emergence of university education offered an opportunity for the state to further regulate medical care. In 1231, Frederick II, the Holy Roman Emperor, enacted a set of laws codifying medication education standards and physicians' licensing; the law required a would-be physician to follow a curriculum of three years of logic, five years of medical study, with special emphasis on reading Hippocrates and Galen, and then a year of apprenticeship. Other universities adopted similar curriculums, although some were shorter; it all looks very modern.* University-educated physicians were able to make diagnoses, prescribe and make up remedies and drugs, and perform surgeries.

In theory, then, a physician in Europe legally couldn't practise without having attained a degree, though enforcement was unevenly applied. This didn't mean, however, that authorities didn't bother at all, as the case of Parisian medical practitioner Jacoba Felicie demonstrates. In 1322, Felicie was brought before a Paris court on charges that she was practising medicine without a license. Felicie, who defended herself, argued that the law was in place to protect people from charlatans, not from skilled practitioners who just didn't happen to have a university education. The Dean and Faculty of Medicine at the University of Paris, which acted as prosecutors, didn't deny that Felicie was a skilled physician and had even cured some patients who university-educated physicians hadn't been able to help. But, they said, she hadn't read the texts. She didn't know her Galen or her Hippocrates. She wasn't *educated.* The court sided with the university and Felicie was prohibited, on pain of imprisonment, from practising medicine in Paris henceforth.

The status and position of medical professionals was increasingly codified and defended by the state and by universities. By the time Chaucer described The Doctor, another pilgrim on his way to Canterbury, in 1380, 'physician' was an established profession – and a wealthy one (well, in some contexts), if his 'blood-red garments,

* Or maybe not entirely modern: in the 1500s, physicians across Europe were required by law to calculate the position of the moon before undertaking any complicated medical procedures, such as bloodletting.

slashed with bluish-gray/ And lined with taffeta' are any indication.* The formation of societies and colleges further concentrated medical power in the few: in the early 1500s, a small group of British physicians successfully lobbied Henry VIII to establish the Royal College of Physicians in 1518. The College, whose powers were expanded to cover all of England in 1523, became the licensing body for medical practice in the country.

However, these university-educated physicians, of which there were only a handful, were not ministering to the general population; for the most part, they served people at the highest strata of medieval and Renaissance society, which also explains how they could press for legal reforms that benefited them. For the vast majority of people, medical care was provided by religious orders (which were prohibited by papal edict from performing surgery and which largely provided palliative care); by surgeons, who doubled as barbers and tooth-pullers; by apothecaries, who, like modern pharmacists, dispensed advice and drugs; by travelling healers who specialised in trades, such as bone-setting or urine analysis (the 'piss prophets'); or by family members, usually women, with midwives taking on the responsibility of birthing. By the 17th and 18th centuries, the medical marketplace was expanding – more apothecaries, more physicians, more street-corner cure peddlers.

Even as medicine was becoming an increasingly formal profession with authority and standing in society, as a practice, it was deeply divided. Universities taught 'heroic' methods of treatment such as bloodletting, purging, and removal of useful organs, as well as medicines that made liberal use of heavy metals, but practitioners and educators rarely agreed even on what they were treating. Internal discord was the case across the developing nations. The first American school of medicine opened in 1765 at the University of Pennsylvania; by the early 1800s, university-based schools were joined by for-profit medical and nursing schools. Doctors who had undertaken a formal medical education – learning those heroic methods – were referred to as the 'allopaths', but there was division in just how that education

---

* Chaucer's Doctor does love his wealth: 'Gold stimulates the heart, or so we're told./ He therefore had a special love of gold.' Oh, snap.

should be applied. And as in Europe, the popular practice of healthcare in 19th-century America was all over the place: DIY healthcare movements, such as the homeopathic Eclectics and, by now, the rapidly growing patent-medicine industry, were also competing for legitimacy in the eyes of government and for the trust of the public.

Compounding this was the lack of consistent governmental regulatory oversight of medical professionals, regardless of what they practised – for example, New Jersey had required would-be physicians to take a licensing exam since 1772, but other colonies and then states didn't. Some states might have had licensing requirements but didn't outline proficiency standards or institute penalties for those who practised without a licence. The biggest and most pressing issue – the one that would decide the fate of medical practice in America – was who would decide which schools were 'legitimate' and qualified to offer a medical degree.

The allopaths, most of whom were white men from wealthier families who could afford the formal medical schools, were deeply invested in increasing the status and authority of their kind of medicine. Their first step mimicked the formalising of medical professions under societies and guilds that had occurred in Europe – creating a unified body under which issues of training, licensing and practice could be hashed out into a credentialing programme. The American Medical Association (AMA) was founded in 1847, with a majority of members belonging to the allopath camp.

The AMA helped sympathetic lawmakers design legislation that would give more oversight power (specifically, deciding which schools were qualified to grant degrees) to individual states Boards of Medical Examiners. These boards were mostly occupied by, you guessed it, members of the AMA. The authority of the Medical Board of Examiners was confirmed in 1888, when Frank Dent, a physician educated at an Eclectic school, sued the state of West Virginia after being convicted of – and fined for – practising medicine without a licence. The case went to the US Supreme Court, which upheld West Virginia's decision that only state-approved schools could grant degrees.

This was a powerful message to all non-licensed and lay practitioners that there was only one kind of 'real' doctor, and only real doctors got to decide who was a real doctor; pharmacists, osteopaths, midwives, homeopaths, patent medicine sellers were all treated with the same kind of disdain. The systematic elimination of the competition, both in North America and Europe, has been well documented and explored by modern medical historians. 'In the drive for the consolidation of professional authority, physicians in the nineteenth century sought to exert control over all associated occupations, and weaken the influence of competitors,' wrote Dr Daniel Malleck in a 2004 article published in *Medical History*. 'First, doctors could and wanted to function without the interference of others in the health industry. Second, doctors sought to enforce a power structure that placed them at the top with all the other health care occupations beneath them, dependent upon the activities of the physicians to maintain their livelihoods.'

At the open of the 20th century in the US, sweeping changes introduced by educational theorist Abraham Flexner, and endorsed by the AMA, resulted in the closing of all but 81 medical training programmes by 1922, down from 162 in 1906. His recommendations were sensible: doctors must be paid, and paid well, no more bartering for care. Education must be evidence-based and built around the scientific method, and every programme must include a year of physical anatomy. Medical training should include four years of undergraduate education, four years of medical school, and three to eight years of internship and residency; once they've completed that, would-be doctors must then pass a state licence exam. These were necessary reforms that nevertheless further concentrated medical authority in the people who could afford it. Similar patterns emerged in Europe. The training and practice of medicine was then enforced not only by professional association but also by the governments that backed them.

The concentration of medical authority in doctors also established the dominance of the biomedical model. Science had, since Descartes, drifted towards reductionism: bodies were reducible to their constituent parts, like a machine with its cogs and gears; illness, disease, injury, pain were all the result of a mechanical fault. The concurrent expanding studies of mind – such as behaviourism, psychiatry, psychology – explained the why of human behaviour, but it was largely understood that mind and body were separate. It was far easier, not to mention rational, to think of pain as resulting from the stimulation of nerves and call it a day.

Some researchers and scientists were cognisant of what was missing in this model. Ronald Melzack, one half of the team that proposed the hugely influential gate control theory of pain mentioned in Chapter 1, wrote in 1968 that the historic focus on pain as a sensory issue meant other potentially useful treatments were ignored: 'We believe that the complex sequences of behavior that characterise pain are determined by sensory, motivational, and cognitive processes acting on motor mechanisms. The therapeutic implications of the model should be obvious; but because of the historical emphasis on the sensory dimension of pain, they are not obvious at all. The surgical and pharmacological attack on pain might well profit by redirecting thinking toward the neglected and almost forgotten contributions of motivational and cognitive processes.'

His advice was ignored. And as authority over the body solidified within the medical profession, the individual who possessed that body became less important. With each technological advancement, each strategic licensing win, each educational reform, the patient's voice became less and less audible, while doctors' voices – mostly male, white, Christian, heterosexual and able-bodied – grew ever louder. And if they said it wasn't pain, well, then it wasn't pain.

When pain isn't painful enough

By the mid-20th century, medical doctors, increasingly valorised in Western society in film and TV, were the acknowledged authority on pain. But there were other stakeholders, too – as the state and, increasingly, pharmaceutical and insurance companies became more involved in the provision of healthcare, the legitimacy of pain became a vital question. These stakeholders demanded a quantitative, objective approach to pain, focusing on what could be observed or measured. They were the ones to determine how pain would be treated, whose pain mattered and who would cover the costs.

Perhaps the biggest impact of this love affair with the biomedical model was on people dealing with what we now call chronic pain. At the moment, around 20 per cent of the global population is suffering from chronic pain, a figure that we can reasonably assume is only going to rise – for one thing, living longer* means we are party to our bodies' slow-motion disintegration in ways that we might not have been in the past. But the problem of chronic pain is by no means a new one, nor is the question of how society should or should not recognise it. During the 20th century, amidst the devastation of two world wars that left millions in pain and as the role of the state was dramatically expanding, the question of what was 'real' pain and what it was worth became vitally important.

In the UK, the Labour party's 1945 landslide victory paved the way for the founding of the National Health Service in 1948. The NHS replaced the previous system of private care and municipal and charity programmes, which wasn't really a system at all, with basic taxpayer-funded care that would be free at the point of service to everyone. At the same time, the US was poised to pursue a similar approach. The Great Depression had prompted a flurry of legislation aimed at getting the country back on its feet through the creation of a federal safety net, including the founding of the Department of Veterans Affairs in 1930 and the Social Security Act of 1935, which guaranteed pensions to all citizens. In 1948, however, though Congress approved parts of President Harry Truman's comprehensive social benefits programme, his universal healthcare plan failed. After 20 years of leadership under Democratic presidents, the culture in Washington was moving towards the right, swayed by arguments of government overreach and excessive expansion of entitlements, and the fear of socialism and communism. In 1952, America elected a Republican president – Dwight Eisenhower, a former general who declared that men were deluded to think that life is free of pain, sacrifice or suffering.

Though political appetite for government aid had waned, the end of the First World War saw the Department of Veterans Affairs struggle

* Well, some of us. The death rate in the US, which dropped year on year throughout the 20th century, has reversed course in the 21st century. According to a major report published in *JAMA* in November 2019, life expectancy in the US increased from 69.9 to 78.9 years between 1956 and 2016, but declined for the last three years. Between 2010 and 2017, the latest year data is available, midlife mortality increased across racial and socioeconomic groups, driven by drug overdoses, alcoholism, suicide and organ systems diseases. And no doubt, the COVID-19 pandemic will also impact our mortality rates..

under the weight of servicemen's needs. Pain was at the centre of political debate, and the attitudes of those in power towards those who were suffering hadn't historically been forgiving. In 1931, a young US soldier who had served three tours of duty on First World War battlefields, been gassed and contracted syphilis at a time when there was no cure or effective treatment, was denied disability benefits under his war insurance policy. By the time his case had reached the Second Circuit Court of Appeals, he'd been dead for six years. But the 1937 decision against him, authored by Judge Learned Hand, established a standard for pain under the law: 'A man may have to endure discomfort or pain and not be totally disabled; much of the best work of life goes on under such disabilities.'

Eisenhower 'would not allow mere sympathy to drive government expansion' but neither could he fail to respond to the demand from veterans – a massive group that, along with their dependants, made up almost half the country – who argued that any attempt to roll back disability payments was a breach of contract made when they agreed to go into service. But many physicians didn't believe that pain claims alone deserved compensation. In 1955, the Commission on Veterans' Pensions asked 153 physicians whether soldiers suffering from pain should be compensated. Although 56 per cent said they should, 41 per cent disagreed. In a comment on the survey, one doctor cited the research of Second World War battlefield medic Henry Knowles Beecher, who found that as many as 75 per cent of severely wounded soldiers didn't request pain relief at the time of injury.* Why then, he wondered, would a man refuse relief on the battlefield, only to come home and ask for *money*?

At the same time, Eisenhower also could not ignore the increasingly loud voices of disability groups, the poor and the aging general

---

* Beecher's research, meanwhile, led him to believe that pain was a much more complex experience than most people understood, that it was specific to the individual and to the context, and that the wound itself was not the same as the perception of pain. Beecher also suggested that the reason soldiers could be cheerful about their painful wounds was that 'his wound suddenly releases him from an exceedingly dangerous environment, one filled with fatigue, discomfort, anxiety, fear and real danger of death, and gives him a ticket to the safety of the hospital.'

population. The AMA was, somewhat bizarrely, on the side of those who argued against more aid, declaring that expanding benefits for veterans would require the expansion of benefits for everyone, the result being socialism and the 'coddling' of the American public. The AMA's real fear, however, was that expanding entitlement programmes would mean the government, not the medical establishment, would be the authority on medical care.

The debate over what the US owed its veterans had morphed into a discussion of what the US owed people in pain generally. For a number of physicians, that wasn't much: some saw chronic pain and people who complained of it as, wrote Keith Wailoo in *Pain: A Political History*, 'a tainted symbol of much that was wrong with society.' But the need among veterans and voters was undeniable. In 1956, Eisenhower reluctantly signed into law the Social Security Disability Insurance (SSDI) programme, which expanded benefits and entitlement to all qualifying disabled Americans – including, critics complained, people who didn't deserve it. Other entitlement programmes, including Medicare, which covered low-income individuals, and Medicare for the elderly, followed soon after.

From the mid-20th century on, distinguishing 'true pain' from fraud and malingering was, as Wailoo characterises, a 'flash point of political controversy'. Differing opinions over the value of pain mapped to ideological extremes. In simple terms, liberals advocated expanding social services, while conservatives worried that 'learned helplessness' would chip away at Americans' social vigour and hardiness, and really, what kind of impression would that give to geopolitical rivals? 'Pain became a signifier with potent political and psychological meanings,' wrote Wailoo, and purging welfare and SSDI rolls was a frequent Republican campaign promise.

Whoever was at the political reins, the adjudication of pain expanded into an industry over the next few decades, encompassing pharmaceutical and health insurance companies, lawyers, medical experts, judges and elected officials. In the 1990s, the field of 'forensic neuropsychology' was born to meet the legal demand for 'symptom validity' assessments and determinations of 'malingering'. The field relied on emerging science about the relationships between brains and behaviour; if you feel like there is a lot of room for junk science analysis and bias in that, well, you wouldn't be wrong. And it all contributed to a pervasive distrust of people in pain across multiple spheres. Writing for a chapter in the 2018 book *Pain Neuroethics and Bioethics*, Amanda Pustilnik noted that legal and medical authorities describe almost all pain claims as fraudulent, or declare that they almost always result from personality disorders. 'At the risk of apparent hyperbole,' she writes, 'the hostility among these decision makers and experts to pain claimants rivals suspicion in the last century towards rape claimants.'

From patient to consumer

Not surprisingly, patient discontent rose throughout the second half of the 20th century. The 1960s and 1970s had ushered in an era of social activism, underpinned by a growing distrust of government and really any institution that wielded too paternalistic an authority – including medicine. As medical expertise had become increasingly specialised and arcane to the general public, physicians more and more seemed to dismiss patient input. Shifting attitudes towards the professionals charged with keeping people alive were reflected in a new crop of books, films and TV shows, such as *M*A*S*H*, which depicted doctors as having their own problems and not always making the right decisions.

At the same time, a series of high-profile stories highlighted doctors' abuse of their power; in 1964, for example, it emerged that researchers at a hospital in New York had injected live cancer cells into patients without their knowledge – and they weren't really even sorry about it. But there was far worse to come. In 1972, any trust built between medical authorities and people of colour was destroyed when the public learned about the Tuskegee Syphilis Study. The study had been commissioned in 1932 by the US Public Health Service to study the progression of the disease. Black men in Alabama were recruited and enrolled under the guise of providing free treatment for 'bad blood', a local term that described general ailments including anaemia and fatigue. Those men who had syphilis were not informed that they did, nor were they given any effective treatment for it, even after penicillin, which cures the infection, was introduced in 1947. The study was only stopped in 1972, after the first news articles began appearing.

In addition to the men who suffered needlessly, victims included the 40 wives who contracted the disease, and 19 children born with congenital syphilis.

Against a backdrop of public debates on medical ethics and civil rights activism, patients' rights groups were gaining ground. They found they were most effective when adopting the rhetoric of consumer rights. For example, in 1975 a consortium of several consumer rights groups successfully petitioned the Food and Drug Administration (FDA) to adopt new requirements for what they called 'patient package inserts' (PPIs). These inserts outlined all relevant drug information and theoretically ensured that patients could make an informed decision on their own. The success of PPIs demonstrated that doctors were not the sole authority able to properly inform patients, and that patients' voices were amplified when they broadcast through the consumer rights platform.

But if patients wanted more say, then pharmaceutical companies were happy to give them something to say. The patient-as-consumer model was useful in the short term, but it also made patients justifiable targets for advertising. From the 1980s, pharmaceutical companies began spending more and more on their direct-to-consumer advertising campaigns, exploding from $55 million in 1991 to $363 million in 1995. Most of this spending was happening in the US; New Zealand is the only other developed nation that allows direct-to-consumer advertising of prescription drugs. Patients now had power – the power to ask their doctors for drugs by name. The consequence? Skyrocketing rates of prescriptions for powerful drugs and an opioid epidemic.

Other concurrent efforts to support patients in pain may also have made things worse. Back in the 1990s, the American Pain Society launched a campaign labelling pain the 'fifth vital sign'. Pain, they believed, was as significant a measure of well-being as blood pressure and heart rate, but was undertreated and unexamined in the clinical context. Treating pain as the fifth vital sign would force practitioners to be mindful of their patients' pain. In 2001, the Joint Commission, the US non-profit that accredits more than 21,000 healthcare programmes and organisations in the country, adopted the fifth vital sign and soon 'pain scales' – numerical ratings corresponding to pain – began popping up in hospitals and doctors' offices across the country; the UK followed suit soon after.

But 15 years after it was widely adopted, pain as the fifth vital sign came under fire from doctors, addiction specialists and others. Their concern was twofold: first, technically, pain is not a vital sign and it can't appear on a beeping monitor as a blinking green number; second, and more importantly, in a marketplace where the lines between patient and consumer are blurred, asking patients about their pain experience introduced the idea that good care was associated with reduced pain. A University of Colorado study from 2018, for example, found that physicians who were incentivised by patient satisfaction reports prescribed opioids for acute pain, cancer pain and chronic non-malignant pain at a rate of three times more than doctors who were not similarly incentivised. The researchers cautiously concluded that, 'Efforts to improve patient satisfaction may have potentially untoward effects on providers' opioid prescribing behaviors.' The patient-as-consumer model both encourages doctors to medicate patients through pain, in the hope of increasing patient satisfaction, and fosters unrealistic expectations among patients that pain, if not entirely avoidable, is at least manageable with drugs alone. In effect, the promise of drugs removed the patient from taking an active role in managing their own pain.

Of course, this isn't a complete picture of all the forces at work in individual healthcare decisions, especially in the US where health insurance companies are the ultimate gatekeepers: they decide which are 'real' conditions and which treatments are covered, and their logic is not always clear. Jacob Gross and Deborah Gordon, of the University of Washington Department of Anesthesiology and Pain Medicine, wrote in an article for the *American Journal of Public Health* in 2019, 'Despite a vast body of evidence of the efficacy of many nonopioid, interdisciplinary care, and self-management interventions, payers often cover more expensive medical interventions even when evidence reveals little benefit.'

Meanwhile, there was another seismic rumble in the healthcare landscape: the internet. The internet might have been invented for the free dissemination of information and, of course, sex, but close on sex's heels was health. One poll from March 2003 found that

66 per cent of internet users said they went online to find information about health; given that this was in the period before the internet's complete domination of most aspects of our lives, that's a lot of people. By 2019 that figure was up to 80 per cent, according to a Pew Research poll, although even that figure seems low.

The internet enabled people to ask their most embarrassing questions without fear of judgement, while at the same time giving them a sense of control and agency. Of course, the internet is the double-edgiest of double-edged swords, just as adept at doling out dangerous misinformation as it is at providing anything of use at all. However, that patients were now able to acquire information that had previously belonged to the healthcare provider alone introduced the potential for a shift in the balance of power, as well as a more collaborative approach to the patient's care. By 2010, for example, researchers were recommending that health educators advise cancer patients to turn to the internet for information, after finding that higher levels of internet use among cancer patients made them want to be more active participants in medical decision-making, regardless of education level.

All of these stakeholders and voices that make up the modern medical industrial complex – the doctor, the state, the pharmaceutical and health insurance companies, the internet – make for a loud, crowded examination room. So loud and so crowded, in fact, that it doesn't leave much room for the person actually suffering.

How (not) to be heard

In November 2017, Jude Taylor* had a hysterectomy. She was 30 years old. Surgery was a relief: by then, she had spent almost half her life waiting for, suffering through or recovering from bouts of immense pain caused by endometriosis; a hysterectomy had the greatest chance of relieving that pain and she had spent more than four years asking for one, over the repeated objections and outright dismissal of her doctors. Her hysterectomy meant that finally – *finally* – she had been heard.

---

* Not her real name; names have been changed to respect the privacy of those in this narrative.

Endometriosis is a deeply vexing and widespread chronic pain condition. In it, cells similar to those that grow inside the uterus – the endometrium – begin to grow outside of it, typically in the ovaries, fallopian tubes and pelvis. In some more extreme cases, the tissue can spread beyond the uterine area, twisting around the intestines. The condition was first identified in the 1920s, and though we know that it is heritable, its exact cause is unknown. And despite the fact that as much as 10 per cent of ovulating women suffer from endometriosis and that it can cause infertility for about 37 per cent of those women, it is monstrously difficult to get a diagnosis of endometriosis at all – studies have found that women can wait up to 11 years from onset of symptoms before being 'officially' diagnosed. If you are a Black woman, it's even harder to get that diagnosis – research demonstrates that Black women are half as likely as white women to be diagnosed with the condition, although there is no difference in rates of endometriosis between Black and white women being treated for infertility.

This is in part because the primary symptom of endometriosis is pain during menstruation – a pain so common and frequently dismissed by school nurses and general practitioners, parents and friends and society in general that most of us simply take its existence as fact. Taylor's extremely painful periods started during adolescence; people around her told her, and she thought to herself, 'Oh, this sucks, you have really hard periods, but you know, periods are painful.' And when some pain is expected, how much is too much?

Taylor's really hard periods, however, saw her blacking out from the intensity of the pain; throughout high school, she usually stayed home during the first day of her cycle, unable to do anything but lie in bed with a heating pad. Her doctor prescribed birth-control pills, the default treatment for any gynaecological disorder*; her monthly

---

* And it had been from the beginning – the FDA originally approved hormonal contraception in 1960, but only for the treatment of gynaecological disorders. It was not until 1965 that the US Supreme Court, in their ruling *Griswold v. Connecticut*, reversed state and local laws criminalising the use of contraception for prevention of pregnancy, but for married couples only. In 1972, the Supreme Court ruling in *Baird v. Eisenstadt* finally gave all persons the right to the pill, regardless of marital status. In the UK, the NHS offered oral contraception from 1961; however, it only explicitly extended that access to unmarried women in 1967.

pain got better or, at least, less awful. Taylor felt that what would make the biggest difference would be to skip her period altogether, which would have been possible by simply taking the pill every day, rather than the traditional 21 days of hormones followed by seven days of placebo to mimic the 28-day menstrual cycle. But decades of medical custom – not science – disagreed.* When Taylor finally convinced her doctor to prescribe the triphasic pills so she could skip it altogether she was warned, 'We don't know what's going to happen with this, you might get cancer in 10 years.' 'Fine, whatever, I'll take my chances,' she replied.

Taylor's pain improved with the elimination of her periods, but it was still a constant in her life. She managed flare-ups as best she could, alternating heating pads and four ibuprofen every four hours. When that didn't work, she'd take a few more NSAIDS, though she soon learned these were taking a tremendous toll on her other organs.

It wasn't until Taylor, who is white, was in her mid-twenties that anyone talked about endometriosis. While she was in graduate school, earning her degree in clinical psychology, her gynaecologist told her that she likely had the condition but the only way to be completely certain was to do a surgical laparoscopy. Even then, her doctor cautioned that there was no guarantee that the laparoscopy would confirm anything: tissue might be too small to see and, in any case, the amount of growth doesn't map to the experience of pain. Taylor remembers her doctor sketching the ovaries and uterus on a piece of paper while explaining how the tissue could be growing outside the uterus. It seemed the sketch was all the doctor really had to offer – well, that and more birth control. This was not the response Taylor had expected. She had endured 10 years of the standard protocol and was ready for whatever was supposed to come next. Surely, she thought, there must be something next. But all she heard from her doctors was, 'Sorry, we can't see it, so we can't treat it.'

---

* The company that introduced the pill in the late 1950s, Searle, would not participate in any trials that would alter the 'normal' 28-day cycle. There is, however, no evidence that skipping periods altogether is in any way harmful; even so, it wasn't until the 21st century that women would be offered a triphasic pill, allowing them to do that.

By 2013 Taylor's pain had intensified. In addition to her usual abdominal pain, she had started to develop crippling back pain. This, she would later learn, was due to the overgrowth of endometrium throughout her abdominal cavity, knitting her intestines to the back of her abdominal wall. The pain became so intense that she eventually went to a place she really didn't want to go – the emergency department.

In hospital, Taylor was given a dose of Dilaudid, the brand name of the opioid hydromorphone, that momentarily tamed the pain. A day and a half later, she was discharged with a one-day supply of Tramadol, an opioid pain reliever with effects equivalent to morphine for moderate pain but not as effective for severe pain. The doctor told her to come back if her pain became uncontrollable again. It was, she judged, a conservative prescription given the amount of pain she was in, but she wasn't surprised. Like most of us, she had seen the opioid epidemic spiral out of control and her doctors' attitudes – and prescriptions – become more cautious.

Two days later the pain was again unbearable. So she did what her doctor told her to do and returned to the same hospital. Only this time, things didn't go so well. From the outset, the admitting doctor seemed sceptical of Taylor's report of unbearable pain. After the initial consultation, she returned with the strong NSAID Toradol and Valium. Taylor looked at the pills in her hand and thought, 'This is insane.' She knew what the pills meant: 'You're not in pain, you're having anxiety around a small amount of pain and just need to calm down.' She didn't want to take the Valium, but she also knew pushing for something stronger would make things worse. Taylor felt like she had to agree: 'You're right, I'm a crazy hysterical female. I'll take this Valium and then maybe if I still say I'm in a really bad pain after I take this Valium, you'll give me what I need.' Taylor was wrong. She waited for four hours, rocking, sobbing and alone. At one point she thought someone was coming in to check on her, but they only closed her door. But Taylor couldn't leave – if she did, it would be considered leaving against medical advice, a note she definitely didn't want showing up on her chart. 'Now I'm literally trapped inside the pain,' she said. 'And inside the hospital.'

Taylor ran through different scenarios out loud: Was there a note in her chart? Was the admitting doctor, who was a resident, trying to prove herself among peers? Did they really think Taylor was just making it all up? Whatever the reason, she knew she would have to wait for a shift change and a new doctor, one she hoped would believe her. Thankfully, they did, but though she got the drugs she needed, suspicion still hung in the air.

After her experience, Taylor tried to take her health into her own hands: 'I started looking into alternative medicine, Chinese medicine because I just wanted to be able to do something … I wanted to feel in control of something.' She took supplements, did pelvic floor exercises; it wasn't enough. Taylor started getting an injection of the contraceptive Depo-Provera every eight weeks, instead of the standard 12 weeks. Her mood and her sex drive plummeted, while her weight went up.

Taylor needed a solution that would address her pain without requiring hormones – so she started asking for a hysterectomy. She knew she didn't want to become pregnant and she was aware of all the risks, including early menopause; it was all worth it if it meant freedom from pain.

Her medical professionals said no. Over and over and over. 'We don't like to give them to women under 30,' they told her. 'What if you want to carry a baby someday?' The subtext, Taylor said, was that women are not competent to make their own decisions, coupled with 'the insidiousness of the belief that all women will want children and it's a primary component of female happiness.' It was enraging. Taylor knew what she wanted and needed but was being told that some time in the future, she'd change her mind. But even if she did, she said, this was not her doctor's problem: 'It's fine with me, it seems like it's not fine with *you*,' she thought. 'I can't get mad at you for *my* decision!'

In 2017, after moving to Seattle, Taylor finally found a gynaecologist who heard her and agreed to give her a hysterectomy. This doctor was confident he could remove the uterus but leave her ovaries, offsetting some of the risks of early menopause. Taylor was scheduled for what she was told would be about an hour of laparoscopic surgery to remove her uterus through three small incisions. But when she woke up, she knew that had not been the case. 'I woke up with awareness that something had gone … and just in that weird intuitive body sense, I could tell I had an incision.' Taylor was told she had stage-IV endometriosis; the surgeon had had to cut a 4-inch incision to remove the growth found across her organs.

When her doctor explained how severe her condition was, she cried – with relief, not fear. 'I knew it, I fucking knew it. I had been saying for years something is really wrong, it shouldn't feel this way,' she said. After being dismissed and ignored, her voice was finally validated: 'No one would believe me and it just was so gratifying.'

Bias and stigma *hurt*

When healthcare gatekeepers won't listen to patients or won't believe them when they do, patients often have to learn a new language and a way of behaving that will give them the biggest chance of being heard. 'I always took weird pride in the fact that I could describe my pain well when I went in to get care,' explained Taylor, who is now a practising clinical psychotherapist. 'This is a place where there's a certain set of norms – and a language – and these are the things I have to do to make myself perceived well and get the things I need.'

Taylor is acutely aware that her odyssey through the medical industrial complex was further complicated by the bias and stigma directed against her, a member of the LGBTQ+ community suffering from a complex pain condition. Doctors are still the voice of authority on the human body and – just like everyone else – they hold biases, both conscious and not.

Bias against pain that cannot be 'objectively' identified is baked into the medical and social care systems that define and treat that pain. As a group of Australian pain researchers wrote in a 2011 article for *Pain Medicine*, 'Because Western medical practice privileges "objective" evidence of bodily lesion over a subjective claim such as being in pain, the absence of evidence of a "disordered machine" defaults to an inference of a "disturbed mind". When a patient's pain is framed in this way by the clinician, there is little chance for meaningful negotiation between the two parties.' Taylor saw this at first hand. 'I experienced all this, like, weird residual frustration from practitioners over my unruly body,' she said. 'Like I'd done an improper job at being diseased.'

This paradigm fuels stigma against people reporting chronic pain. And that stigma is *huge*. In 2011 the Institute of Medicine's Committee on Relieving Pain in America published its blockbuster report on chronic pain, highlighting testimony from the 2,000 people living with pain, and their loved ones and caregivers. As one person told the committee, 'I have a Masters degree in clinical social work. I have a well-documented illness that explains the cause of my pain. But when my pain flares up and I go to the ER, I'll put on the hospital gown and lose my social status and my identity. I'll become a blank slate for the doctors to project their own biases and prejudices onto.'

People with chronic pain struggle – and have always struggled – to be believed and respected. 'Qualitative evidence indicates [people with chronic pain] do not feel believed by romantic partners, relatives, and friends. They believe practitioners think their pain is exaggerated or imagined. They feel blamed, misled, and even report being dismissed by health care providers,' wrote Lies De Ruddere, a psychologist at Ghent University, and Kenneth Craig, of the Pain Lab in the University of British Columbia's Psychology Department, in the journal *Pain*.

Stigma takes its toll on everyone in different degrees, and research shows that how deeply it cuts is largely determined by how aware the patient is of stigmatising views and the extent to which they apply those views to themselves. Individuals suffering chronic pain *and* stigma might begin to question their own credibility and perception, not to mention their value. And lest we be tempted to think that stigma too is 'all in their heads', it's not. Findings consistently demonstrate that medical professionals trust their own assessments of a patient's pain over the patient's self-report and that observers attribute lower pain values to individuals deemed responsible for their own injury; people with chronic pain or pain with no clear aetiology are perceived as less likeable and less believable; experts, meanwhile, routinely debate the existence of certain chronic pain conditions, despite the millions of people demonstrably suffering them.

'Patients with pain are lightning rods for social blame and stigma,' acknowledged Dr Daniel Carr, president of the American Academy of

Pain Medicine in a 2016 article for *Pain Medicine.* Why? Carr described humans as social animals who want to maintain their group's cohesiveness and stability. 'This stability includes feeding and protecting the group from external assaults. Many packs or herds support their members immediately after injury,' he wrote. 'But with time, as they become less and less able to help feed or protect the group, they turn their backs on them, rejecting and expelling them.' In other words, humans will support the sick and injured to the point that the sick and injured become a burden. And then they won't. But though we are animals and we all have our limitations, we're also gifted with the ability to think about what and why we do things, meaning 'evolution made me do it' doesn't work as an excuse.

Though by no means 'hardwired', our reactions to other people are informed by cultural adaptations. In the last chapter, we talked about pain expression as a kind of social evolution – nociceptive pain is an alarm system, necessary for our survival, as is being able to communicate that pain to others. Which is a problem when it comes to chronic pain. Most people think they know pain when they see it, but evidence suggests that our ability to accurately read pain in others is limited. While the expression of acute pain is often reflexive and automatic, the expression of chronic pain is more often a combination of description and 'guarding' or protective behaviours. Chronic pain expressions do not seem to carry the same legitimacy: for example, the belief that if a person is able to talk about it, well, then the pain must not be that bad. In fact, studies have shown that people who talk about how severely their life has been impacted by pain are perceived as less likeable, as are those who report higher levels of pain severity.

Taylor's experience in the emergency department was also complicated by a new social reality that people in chronic pain are facing. She believes her biggest 'mistake' in returning to the hospital was specifically asking the admitting doctor for Dilaudid, the same drug that had given her relief only two days before. Dilaudid is one of the opioids that people seeking drugs come to emergency departments to ask for explicitly, she later learned; research also demonstrates that physicians also associate asking for drugs by name as drug-seeking behaviour. 'And as far as I can tell from that, it was over for me.'

As the consequences of the opioid abuse epidemic continue to devastate lives and communities, doctors are no longer solely charged with treating pain – they are also tasked with stemming abuse by assessing whether they think the person presenting pain is 'trustworthy'. This assessment is frequently the result of implicit biases and snap judgements based on demographics – ethnicity, age, gender, skin colour, socioeconomic status and attractiveness. Besides being an ineffective way of dealing with real drug addiction, the increased scrutiny and disbelief among medical professionals leaves those seeking pain relief in fear that they will be treated like a criminal, a drug-seeker, and distrusted or dismissed. Sometimes, that fear is realised. That's what happened to Taylor, she thinks: '[The ER doctor] came in, decided I was drug-seeking and then that crossed the whole grapevine and then nobody would pay any attention to me. So they literally just left me there.'

Being labelled a drug-seeker had a demonstrable effect on Taylor's treatment – it left her waiting, alone in a hospital room, terrified and in pain. But that isn't the only bias that Taylor faced or could have faced. As a member of the LGBTQ+ community, Taylor braced herself for the potential of homophobic or ignorant medical professionals with every new medical encounter. She doesn't know if that factored into this experience, but it certainly has done in the past. Discrimination against LGBTQ+ people in healthcare is a global public health concern, especially in places where same-sex relationships are illegal. But even in countries where they are not, prejudice against LGBTQ+ people is real and dangerous. In a representative survey of almost 3,500 US residents, 18 per cent of LGBTQ+ respondents reported avoiding healthcare due to the anticipation of discrimination, and 16 per cent reported experiencing discrimination ranging from overt slurs to more subtle but also powerful microaggressions.

Even beyond deeply entrenched discrimination and prejudice, all of us maintain unfounded biases, most of which we're only barely aware of. But these creep into our assessments of what we think other people are feeling, and if you're a medical care professional, the consequences for patients can be devastating. The last thing anyone wants to deal with when asking for help is a challenge to their right to be treated with dignity and respect. And yet, it happens – a lot.

False beliefs rooted in prejudice and junk science are nothing new; in the 1870s, psychologist Cesare Lombroso conducted a series of experiments measuring the pain threshold of Italian prisoners. Lombroso subjected incarcerated men to a device he called a 'dolorimeter', which delivered electrical stimulus in controlled, measurable amounts, and recorded the point at which subjects said 'ouch'. In this case, the point at which convicted criminals said 'ouch' became an indicator of their inherent criminality, a trait of their 'type'. Lomboro wrote in his 1876 book, *The Criminal Man*, 'Compared with ordinary individuals, the criminal shows greater insensibility to pain as well as to touch. This obtuseness sometimes reaches complete analgesia or total absence of feeling (16 per cent), a phenomenon never encountered in normal persons.'* Lombroso's eugenicist findings were flawed, to say the least, but popular; similar experiments came up with similar 'findings' about the relative pain thresholds of women, Europeans and non-Europeans, and 'deviants'.

Despite their training and despite the nearly 150 years it's been since Lombroso, medical professionals *still* maintain false and racist beliefs about the pain of others: A 2016 University of Virginia study found that 'a substantial number of white laypeople and medical students and residents hold false beliefs about biological differences between Blacks and whites', such as the myth, born from racist ideology, that '[B]lack people's skin is thicker than white people's'. The study found that half of the white medical students surveyed maintained false beliefs; researchers concluded that some white people with medical training 'hold and may use false beliefs about biological differences between Blacks and whites to inform medical judgments, which may contribute to racial disparities in pain assessment and treatment.'

Studies consistently demonstrate that bias, whether based on false beliefs or not, affects how doctors relieve pain. A 1997 study into pain relief for outpatients with recurring or metastatic cancer found that Black and Hispanic people were consistently given inadequate pain

---

* Among Lombroso's other 'findings' was that criminals had a significant tendency to be left-handed, and that women were just 'undeveloped men', and that those 'undeveloped men' who became prostitutes loved orgies and were short and left-handed.

relief – 74 per cent of Hispanic and 59 per cent of Black American patients reporting pain received less than the World Health Organization-recommended dosages of analgesics. That was in 1997, but little has changed since: according to a 2016 study, Black patients are half as likely to be prescribed opioid drugs in an emergency-room setting as white patients, while a 2015 study found that Black children suffering from appendicitis are less likely than other races to receive pain medication, including for severe pain, in the emergency room.

There are so, so many examples of how bias *hurts*. A 2019 study of individuals suffering from ureteral stones – that is, kidney stones that have become lodged in the ureters, the tubes that carry urine from the kidneys to the bladder – found large disparities in how emergency departments treated them. People over the age of 55 were less likely to receive pain medication compared to younger patients; women were less likely than men to see any form of diagnostic imaging to confirm the stones and were also less likely to receive alpha blockers to help facilitate the stone's passing.

Other studies have confirmed what obese patients already feel is true – that medical professionals are judging them and, in fact, some hold vicious opinions about them. Patients who are morbidly obese are most likely to be the targets of derogatory humour from medical students, attending physicians and residents; 24 per cent of nurses reported being 'repulsed' by overweight patients, while 12 per cent said they didn't want to touch obese patients; 31 to 42 per cent of nurses said they'd prefer not to treat obese patients. Studies also demonstrate that obesity is correlated with increased self-reported pain: if you think your obese patient is fat because they 'lack self-control' and that they're in pain because they're fat, how likely are you to provide pain relief?

And *knowing* that you're the victim of bias hurts – perceived injustice makes pain worse. The evidence is consistent. An article published in *Pain* in 2016 reported findings from a study of 139 teens: 'Higher levels of perceived injustice were associated with higher levels of pain intensity, catastrophizing, and functional disability, and with poorer emotional, social, and school functioning.'

Given the bias that pervades many medical interactions – not to mention most *human* interactions – that Taylor's endometriosis was dismissed for years as just 'painful periods' isn't at all surprising. Biases in pain treatment disadvantage women and have for a very long time. In 2001 a landmark paper entitled 'The Girl Who Cried Pain: A Bias Against Women in the Treatment of Pain' in the *Journal of Law, Medicine and Ethics* showed that women are more likely to seek treatment for chronic pain but are also more likely to have their reports disregarded by clinicians. Women reporting the same pains as men in emergency rooms wait longer for analgesics: a widely reported 2008 study found that, after controlling for age, race, triage class and self-reported pain scores, women were still 13 to 25 per cent less likely than men to receive opioid analgesia, and waited longer than men to receive any analgesia at all – a median time of 65 minutes versus the 49 minutes for men.

For much of the 20th century and well before, men dominated the medical and science professions; women's pain, illnesses and diseases – as distinct from men's – were, to some degree, invisible. Though many more women are now in these professions, this is one hangover that's tough to shake: in 2018 women across the world heaved exasperated sighs when dozens of global newspapers gave headline space to a male University College London professor of reproductive health who made an off-hand comment that menstrual cramps can be as painful as having a heart attack. We'd say something snarky about this, but Arwa Mahdawi in the *Guardian* already did it better: 'Women have been saying for centuries that period pain is excruciating. But, I mean, women are prone to being a little hysterical. It's best to take everything they say with a pinch of salt and consult a man for a second opinion … It was literally just a male figure of authority reporting his patients' comments – and yet it made international news.'

We have the will – now we need the way

Taylor thought she was cured after her 2017 hysterectomy, but a year later she noticed a bump on her abdomen. She made an appointment with her primary care doctor, thinking it might be a build-up of scar tissue. An ultrasound revealed it was actually 5 and a half inches of endometrium tissue, winding its way around her ovaries and fallopian tube. Her doctor immediately referred her to a gynaecological oncologist.

'By the time you level up to oncology, apparently, they stop bullshitting you,' Taylor said. This time, the surgeon would be removing all remaining growth, along with her ovaries, fallopian tubes and appendix, just in case – because the risk of complications increases with each surgery, she didn't want anyone having to open Taylor back up again. The gynaecological oncologist also warned Taylor against using hormones to manage the condition post-surgery; the risk of feeding any microscopic remaining endometrium was too high. A host of other health concerns, including the increased risk of heart attack and the loss of bone density, were now also on Taylor's radar. The news was tough to hear but, said Taylor, it was the first time she felt like she was getting all the information.

Two days later, Taylor was in the operating room. When she woke up this time, she had a 5-inch vertical incision intersecting the 4-inch horizontal scar from just a year before.

At the time of our interview, a year to the month after her last surgery, Taylor had just learned she had 'graduated' from the gynaecological oncologist. It's making her feel a lot of complicated emotions she's still trying to identify and process. 'I just resigned myself to my life spent in gynaecology,' she said. 'So I don't even know how to describe these complicated feelings that arise with this announcement that "You never have to come back here", where they're going to weigh me on a scale with like a little pink flower-trimmed sign that says "Chocolate will solve all problems".' One thing she does know, however, is that she will not miss that sign.

Taylor's pain was *her* pain, but by not listening to her, doctors tried to make it *theirs*. Not because doctors are sadistic jerks who want their patients to suffer; usually, it's the exact opposite – they want their patients to *not* be in pain. But the limitations of the biomedical model aren't just damaging to people whose pain cannot be accounted for under it; it also leaves doctors trying to treat pain with one hand tied behind their backs. Most doctors, believe it or not, are not sufficiently trained to manage pain and, when they don't have the answers they feel they're supposed to have, they get frustrated. And sometimes, as Taylor observed, that frustration is projected onto the patient.

'Doctors hate pain. Let me count the ways. We hate it because we are (mostly) kindhearted and hate to see people suffer. We hate it because it is invisible, cannot be measured or monitored, and varies wildly and unpredictably from person to person. We hate it because it can drag us closer to the perilous zones of illegal practice than any other complaint,' wrote physician Dr Abigail Zuger in a 2013 Well column for the *New York Times*. 'And we hate it most of all because unless we specifically seek out training in how to manage pain, we get virtually none at all, and wind up flying over all kinds of scary territory absolutely solo, without a map or a net.'

That's right: despite the fact that they are the acknowledged experts on the legitimacy and treatment of pain, doctors get very little education in the matter. A 2011 survey of 117 medical schools in the US and Canada found that, though most schools mandated a session on pain, these were typically presented as part of general required courses; researchers concluded that 'pain education for North American medical students is limited, variable, and often fragmentary.' The situation isn't any better in Europe: a 2013 survey of all medical schools in 15 representative European countries found that the overwhelming majority taught pain only within compulsory but non-pain-specific modules and concluded that, 'Documented pain teaching in many European medical schools falls far short of what might be expected given the prevalence and public health burden of pain.'

The scope of this problem is immense. In 2012 doctors Philip A. Pizzo and Noreen M. Clark, co-chairs of the Institute of Medicine's committee on pain in the US, described a nation throwing money hand over fist at a chronic pain problem spiralling out of control, with little to no relief. In an article for the *New England Journal of Medicine*, they wrote:

> Major impediments to relief include patients' limited access to clinicians who are knowledgeable about acute and chronic pain – owing in part to the prevalence of outmoded or unscientific knowledge and attitudes about pain. Fundamental differences in views about pain and its management pervade the medical profession. Some physicians overprescribe medications including opioids, while

> others refuse to prescribe them at all for fear of violating local or state regulations ... Decisions about medical care are also influenced by insurance coverage that may be preferential for injections, infusions, procedures, and surgery over the physical therapy, rehabilitation, or other more comprehensive approaches to pain control that may benefit patients more. Physicians' referral of patients to other health care professionals, including nurses, chiropractors, and practitioners of complementary medicine, and patients' willingness to seek such care, can be influenced by bias, unclear data, and the availability of care. Sadly, many people with chronic pain see physicians as 'poor listeners.'

Part of the current over-reliance on pharmacological interventions is down to the fact that many healthcare professionals simply don't know what else to recommend to their patients in pain. After decades of building physicians' authority, the pressure on them to be able to just fix it is tremendous. A 2016 survey of 4,925 physicians, nurses, physical therapists and midwives at a Swiss university hospital found that the vast majority, 96.1 per cent, agreed that complementary medicine – such as hypnosis, osteopathy, acupuncture, mindfulness-based stress reduction, biofeedback and neural therapy – could be useful in the treatment of chronic pain. But more than half the respondents, at 58.3 per cent, had *never* referred a patient to a complementary medicine practitioner. Overall, a total of 84.3 per cent felt they lacked the knowledge to inform their patients about what was available to them. In other countries, as Pizzo and Clark pointed out, they can't – many US insurance companies don't cover therapies they consider alternative or complementary, despite evidence that incorporating things like mindfulness therapy, yoga and relaxation techniques can help reduce opiate dependency in patients with chronic pain.

For most of the 20th century, advancements in pain treatment were hampered by deeply entrenched, flawed thinking about pain and the concentration of medical authority in a few. The conventional perception of pain as a problem of sensory input, a paradigm constructed more than 350 years ago, might have been useful in the study of the physiology of pain, but it's done a bit more harm than good in the management of the lived experience. It's also resulted in the stigmatising of pain that's 'in your head', both from medical practitioners and, crucially, patients themselves. The concentration of medical authority in middle- to upper-class white men meant pain that didn't fit their definition wasn't pain at all; once the state got on board with this, it also meant the social safety net could have some fairly large holes. It enabled a paternalistic system that prioritised the physician's power over the patient's autonomy, and meant that sticking an ice pick through someone's eye cavity without their consent could be considered a legitimate pain treatment. And it meant, as we'll see, the appropriate treatments for pain were only those that the doctor could dispense.

That meant drugs. Lots and lots of drugs.

CHAPTER THREE

The Cure and the Cause

For many of us, drugs seem to be the only option for dealing with pain, whether it's a headache, a broken arm or a broken heart. And they can be a good option. It's just that the case for pharmaceutical interventions is more mixed than we know and we're using them in place of other therapies that might actually work just as well and come with fewer side effects. Often, drugs are a blunt tool for a delicate problem.

All drugs – but particularly drugs that purport to treat pain – can come with a host of unintended consequences and side effects, only some of which are listed in the fine print. Meanwhile, the horrible irony is that many pain-relieving drugs aren't even doing what they're supposed to be doing: when it comes to treating the most severe forms of chronic pain, for example, opioids perform in some trials worse than placebos.* In fact, there is growing evidence that prolonged opioid use might actually be making some users abnormally sensitive to pain, a phenomenon called 'opioid-induced hyperalgesia'. Though the prevalence of opioid-induced hyperalgesia is unknown, the science seems to indicate that opioid use causes neuroplastic changes in the PNS and CNS, leading to the sensitisation of certain pain pathways.

That's not even touching the serious toll that opioids take on all our systems, resulting in difficulty urinating, nausea, vomiting, constipation and loss of sexual function. No wonder there's mounting evidence that opioids can exacerbate anxiety and depression in people with chronic pain, a population already 15 to 20 per cent more likely to experience negative affective states. One doctor, in an opinion

---

* However, as much as that doesn't speak well of pharmacological analgesics, it does open another avenue for pain relief. Placebo has been anecdotally successful for, well, ever but perhaps it's time for the effect to be introduced more formally in the treatment of pain. More on this later.

piece for the *New York Times* in 2019, described prescribing opioids for non-cancer chronic pain, such as chronic backache and arthritic knees, as 'the worst medical mistake of our era.'

The global struggle with opioids – the towns decimated by addiction, the children being asked to carry Naloxone kits and learn how to reverse an overdose, the deaths – is well documented. But it's part of a larger narrative around pharmaceutical therapies, one in which better living through chemistry has contributed to a dangerous mindset. Over the course of the 19th and 20th centuries, we, as a society, have become less tolerant of the idea of pain. Drugs are a big part of this paradigm – the better our pharmacological interventions got, even if they weren't perfect or useful all of the time, the more we expected of them. Decades of improving pain treatments diminished our tolerance to live with pain of all types; drugs themselves got stronger.

As Fernando Cervero wrote in his 2012 book, *Understanding Pain*, 'As we improve our pain treatments and develop new and better analgesic procedures, we get caught in a vicious circle that makes a pain-free world an impossible goal.' When we start to experience chronic pain – the kind we are increasingly suffering – we expect to be met with something, anything, to take that pain away. Doctors want to help us, so they keep trying with the tools they have, which for pain means mostly drugs. Now, we're in a weird place where we have greater access to more powerful drugs, combined with a diminished tolerance for discomfort and an increased expectation of feeling good. Correlation isn't causation, but it is interesting that the nations with the highest rates of chronic pain, including the US, Australia and the UK, tend to be the nations that rely more heavily on drugs of all stripes for pain relief.

We're pointing an accusatory finger at big pharmaceutical companies and are joining the chorus railing against the over-prescription of opioids – more on that shortly – but as our millennia-long love affair with opium and other drugs demonstrates, humans have always looked to drugs to muffle, stifle or remove the sensation of pain. And good on us! The amount of relief brought to humans and animals alike thanks to drugs is incalculable. Opium is one of the oldest forms of pharmacological pain relief, used so widely and consistently that it's difficult to pinpoint when and where it first came into cultivation or even its native region. The coca leaf, rich in the alkaloid stimulant cocaine, has been central to health practices in Central and South American cultures since as early as the ninth century BC; Incan surgeons chewed coca leaves and dripped their coca-infused saliva into the open incisions of people they were trepanning. Cannabis has been in constant medicinal use for the last 5,000 years, originating in what is now Romania. It also gets a mention in the Ebers Papyrus as an anti-inflammatory, and in 600 BC India's famed physician Sushruta used cannabis vapours to sedate surgical patients. And, just as they are now, these substances were as likely to be used recreationally or ritualistically as they were medically, as likely to be over-used as they were deployed helpfully. Opportunities for harm and for healing were intertwined in the discovery, creation and distribution of chemical compounds from the very beginning – the cure and cause of suffering could be found in a single leaf, pill, powder or bottle.

However, it wasn't until the 19th century that drugs became readily available to the masses as a means of pain management – and, at the same time, became a problem. By the middle of the century, the *New York Times* estimated that as much as 1,000 pounds of opium was being sold over the counter in New York City alone each week. By 1888 opiates – drugs derived from opium – made up 15 per cent of all prescriptions dispensed in Boston pharmacies. In the UK the number of chemists and druggists quadrupled between 1865 and 1905, responding to the increased need and desire for drugs. And then there were the thousands of patent medicines, many of which relied on sham ingredients combined with opium, alcohol and cocaine to effect a 'cure'.

Some of that explosion had to do with the expansion of global trade, the rise in incomes and the growth and regulation of so-called vice economies; some of it had to do with the refinement of certain drugs, including opium and cocaine; and some of it had to do with the adoption of anaesthesia in surgery. All of it had to do with the changing perception of what pain was and whether or not we needed it.

Getting high and going under

Surgery in the early 19th century included the removal of tumours, amputation of diseased limbs, 'lithotomy' or the removal of bladder stones,[*] tracheostomy and more. But the possibilities of surgery were limited by the amount of acute nociceptive pain a patient could withstand. Alcohol and opium were sometimes made available, but these came with drawbacks (a drunken patient might not be a very compliant one).[†] Some anaesthetic gases and substances, such as ether, had been around for hundreds of years, and some scientists had certainly investigated them and their uses. But this was in a time before a standardised scientific method, so results tended to be a bit haphazard. In the years during and after the Enlightenment, as sciences started to harden, experiments were more tightly controlled. To a point.

Nitrous oxide (N_2O), a colourless, flammable gas with a slightly sweet scent, was first synthesised in 1772 by Joseph Priestley, the British chemist who also invented soda water.[‡] But it wasn't until 1799 that someone tried to figure out what breathing nitrous oxide would do to the human body. That someone was the fearless – or foolish – Humphry Davy.

Davy was the eldest son of a Cornish wood-carver and an energetic scientific prodigy who'd learned French from an exiled priest at the age of 14 just so that he could read chemistry books written in the language. Aged 18, Davy found work as the first superintendent of the Pneumatic Institute, a treatment facility located just outside of Bristol that experimented with the idea that certain gases, when inhaled, could cure various illnesses. Davy started experimenting with nitrous oxide by synthesising it and inhaling it himself. When he didn't die – and actually quite enjoyed the experience, much more than his experiments with breathing carbon monoxide – he began administering it to patients, who all seemed to look forward to their 'dose of air' and 'the pleasure it gave them'. Davy also looked forward to his doses of air; at the start of the summer, he'd embarked on a rigorous course of self-experimentation. Davy took detailed notes on the effects of nitrous

---

[*] In the 1740s, William Cheselden was considered one of the best lithotomists in Britain. He enjoyed a 50 per cent mortality rate for perineal lithotomy.

[†] In 2014 Vijay Welch-Young underwent surgery to remove a tumour wrapped around his heart without anaesthesia. He doesn't recommend it.

[‡] Priestley was also driven out of Britain after he made his dissenting views about the monarchy too plain – he set up in the US after a mob tore up his priceless chemical lab, burned an effigy of him and destroyed his house in 1791.

oxide inhalation on his blood pressure, pulse and body temperature as well as the effect on his conscious state: the gas, he wrote, has 'made me dance about the laboratory as a madman.' Davy's experiments reached a high point on Boxing Day 1799, when he walked into a purpose-built sealed box and ordered a friend and physician to keep pumping it full of nitrous oxide until he passed out. After an hour and 15 minutes, Davy was still conscious. He left the box and huffed a further 20 quarts of nitrous oxide from oiled silk bags he'd had made for the purpose. Davy was absolutely off his face: 'Nothing exists but thoughts!' he cried.

In 1800, Davy published his work on nitrous oxide, *Researches, Chemical and Philosophical; Chiefly Concerning Nitrous Oxide, or Dephlogisticated Nitrous Air, and Its Respiration*, detailing his personal experiences with it and describing its qualities. In it, Davy made several mentions of the fact that nitrous oxide seemed to inhibit pain, concluding, 'As nitrous oxide in its extensive operation appears capable of destroying physical pain, it may probably be used with advantage during surgical operations in which no great effusion of blood takes place.'

The work was a blockbuster success and it paved the way for Davy's celebrated career as a chemist; he became president of the Royal Society and earned a knighthood, and his self-experimentation was seen as proof of his dedication to science. But the gas that put the wind in his professional sails was largely ignored by medical science. That it could 'destroy' pain seemed to be of little interest, outside of novelty; the discovery that it induced a sense of euphoria and giddiness was far more interesting. 'Laughing gas' parties were hugely fashionable (for those who could afford them), and by the 1820s nitrous oxide was a staple of variety acts on stages across Britain and America. The act ran much in the way that comedy hypnotist shows do now: an actor playing a doctor would invite audience members on stage to inhale the gas; the entertainment was watching the participant stumble around, instantly intoxicated.

The lack of interest in nitrous oxide as a surgical aid could represent a scientific blind spot on the part of Davy and the scientists who followed, however, there were some issues with nitrous oxide's use as an anaesthetic. For one thing, the euphoria and hilarity it induced in a subject wasn't exactly conducive to surgery – no one wants to cut open someone in the grips of a laughing mania. But for another thing, it underscores the place and role of pain in Western society at the opening of the 19th century. Pain was, by and large, a necessity – regrettable and unpleasant but necessary, to be endured rather than stifled. Davy himself stated that 'a firm mind might endure in silence any degree of pain', showing the supremacy of 'mind over matter'; religion, meanwhile, still maintained that physical pain and suffering had a value, if not in this world then the next. Pain was God's will, and actively seeking to avoid it would risk God's wrath.

There was also a medical interest in keeping pain in the operating theatre. In 18th- and 19th-century medical practice, certain kinds of pain were seen as a necessary indicator of vitality. Early 19th-century doctors objected to anaesthesia because it was perceived as depressing the body's natural inclination towards activity; anything that put the patient in such a deep sleep that being cut open didn't wake them couldn't have been healthy. 'Pain was understood to be, by and large, a benefit, it was like the body's safety net,' said Dr Stephanie Snow, professor of medical history at the University of Manchester and author of *Blessed Days of Anaesthesia*, a history of anaesthesia, in an interview with Linda in 2015. 'If you were having an operation and you felt pain, that was actually a good thing because the pain was a trigger to the body to maintain its vitality. If you had a patient on the operating table and they sort of cried out and writhed in pain, that was good, it meant that they were alive.'

In the space of about 50 years, however, this perception of pain – as both useful and endurable – would change dramatically. 'From the 1820s onwards, you get the sense that causing anything pain, even animals, is against the basic impulse of a civilised society ... Taking away pain is actually a blessing,' explained Snow. Members of 'civilised society', as early 19th century people wanted to be, liked appearing as if they weren't in pain, as if they had severed that link to their baser, animalistic selves; this tallied nicely with the late-Georgian and Victorian impulse towards self-control (how laughing-gas parties fit into this paradigm, well, let's just say there were some social contradictions). 'The act of crying out, the expression of physical feeling, was seen as weak self-control,' Snow said. In this context, that opium and morphine were as popular as they were makes more sense – in some ways, they were the perfect drug for an era obsessed with detachment and calm. As pharmacological pain relief became more common and more popular, pain was no longer seen as something we needed. Meanwhile, the moral value that pain once had was being chipped away as the power of the church waned. Pain didn't have a purpose or a place in modern times.

As the perception of pain continued to change, some scientists began to explore using nitrous oxide and substances like it, including ether, specifically to blunt acute pain. Dentists were among the first people to start experimenting with pain relief during procedures, which makes a lot of sense: tooth extractions, though extremely painful, were unlikely to be fatal or occasion a lot of blood loss. A good dentist could perform the extraction as fast and as painlessly as possible – that was how they marketed themselves. And if they could truly offer painless tooth extraction, then they were really in business. In 1845, dentist Horace Wells had attempted to usher in what he bravely called a 'new era for tooth-pulling' by demonstrating the use of nitrous oxide as an anaesthetic in front of a group of doctors at Massachusetts General Hospital in Boston. Wells, however, had pulled the bag containing the gas from the subject's face too soon and the man moaned in pain during the extraction. The whole experiment was derided as a 'humbug'; both nitrous oxide's and Wells's reputation was shredded.*

But that wasn't the end of anaesthesia. Where nitrous oxide – or rather, Wells – had failed, ether would succeed. Ether ($CH_3CH_2 - O - CH_2CH_3$, frequently abbreviated to Et_2O) is a liquid compound that is typically produced by distilling ethanol and sulphuric acid. It is similar to nitrous oxide in its effects, only much stronger and more

* Wells's story is tragic. After the debacle at Mass General, Wells threw himself into experimenting with inhaling chloroform, which had been invented in the US in 1831. But Wells went on a four-week bender in New York City and, in a manic fit, he rushed out onto Broadway and threw acid on two women; he was arrested. While in jail on 24 January 1848, he took his own life by slicing his femoral artery with the razor from his shaving kit. But not, it should be noted, before he'd first managed to procure some chloroform to inhale.

flammable; it boils at 35°C, rendering it a gas fairly easily. The early history of the compound isn't entirely clear – it was either synthesised in the eighth century or in 1275 – but it was definitely synthesised in 1540 by chemist Valerius Cordus, who discovered it by pouring sulphuric acid on the strongest distilled booze he could get his hands on. The resulting compound, called 'sweet oil of vitriol', was found to have medicinal properties; contemporary Swiss scientist Parcelus used it to put animals to sleep (in the actual sense, not the euphemistic one). Called 'ether' from 1729 by chemist August Sigmund Frobenius in a treatise about its properties, it was in use as a vapour from the late 1700s; Edinburgh physician William Cullen recommended breathing ether as a treatment for headache. Up through the 19th century, vaporised ether was used as a treatment for illnesses such as scurvy, croup, asthma and catarrh, not because it actually did anything to manage those illnesses but because it did anything at all.

It was also fun. People across Europe regularly drank it – neat, rinsing their mouths out with cold water before and after – to get around temperance injunctions against alcohol. 'Ether frolics', similar to laughing-gas parties, were also popular in Europe and in the southern American states, where distilling grain alcohol into ethanol was (and is) a cultural pastime. But though people had long observed its anaesthetic qualities, it wasn't until the 1840s that ether was introduced into the operating theatre.

On 16 October 1846, William T.G. Morton rendered Edward Abbott insensible under the ether dome at Massachusetts General Hospital. Morton was a dentist who had previously used diethyl ether vapours to perform painless tooth extractions, but a lot was riding on this demonstration – after all, this was just a year and a half after Wells's disastrous showing in the same venue. Many doctors and surgeons believed that focusing on pain in surgery would terrify some patients and prevent them from getting necessary procedures. Many more believed that painless surgery was impossible; Valentine Mott, a well-respected New York surgeon, said that very same year that anaesthesia in surgery was a 'chimera that we can no longer pursue in our times.' Plus, this wasn't just a tooth extraction – Morton was attempting to anaesthetise a man who would be having a tumour removed from the left side of his jaw. But where Wells's nerves had got the better of him, Morton's natural showmanship shone, charming not only the medical men in attendance in the operating theatre, but also the patient himself, who told him, 'I feel confident that you will do precisely as you tell me.'

Morton applied the ether mask, a device he'd made himself, to Abbott's face. Within three or four minutes, Abbott was out, reclining in a chair. Morton gave the signal to surgeon-in-chief John Collins Warren to begin the operation. And even as the scalpel sliced into his skin, Abbott made no sound or movement. He remained insensible throughout the operation, although towards the end, as the ether began to wear off, he seemed to move his limbs uncomfortably and cry out – but once he was awake and talking again, Abbott was insistent that he'd felt no pain, barely a scratch. 'Gentleman,' Warren told his assembled colleagues, 'this is no humbug.'*

By the end of the year, ether's qualities were not only heralded across the US but in the UK as well. On 21 December 1846, ether was demonstrated at University College Hospital in London, during the amputation of a man's leg. Robert Liston, professor of surgery at the hospital, reportedly declared, 'Gentlemen! This Yankee dodge beats mesmerism hollow!' Later that night, Snow wrote in *Blessed Days of Anaesthesia*, he hosted a celebratory dinner party during which he let one of his guests try out ether (as far as we know, that demonstration didn't include an amputation as well).

Within months of ether's successful debut at Massachusetts General Hospital, the drug was also being used to relieve pain in childbirth, although uptake in that area was contentious and somewhat slow. By 1852 – the year Congress was considering what, if anything, to reward

---

* Morton's success was tainted, however, by what would become a decades-long struggle – that would eventually reach Congress – to reap the financial rewards of his discovery. In the first place, there was some question as to whether he really did discover ether's qualities as an anaesthetic; a surgeon in Georgia, for example, had used ether to pacify a patient who needed a cyst removed several years earlier. Meanwhile, Morton's landlord, also a dentist, claimed he had first given Morton the idea to use ether and so he ought to get a cut. Even Wells threw his hat into the ring, claiming it was his experiments with nitrous oxide that started the whole thing and that he'd 'leave it to the public to decide to whom belongs the honor of discovery.'

Morton for his 'discovery' of ether – the gas was in use throughout the American military as well as in surgeries and childbirth beds across Europe. This was despite the risks associated with using ether, which seemed to affect patients variably. In 1854, for example, a 'beautiful young girl, 18 years old' died during an operation to remove a tumour from her neck after being anaesthetised by ether. She wasn't the only person to slip into an ether sleep and never wake up: reports of individuals dying under ether were increasingly in the headlines. At the same time, other people seemed to breathe the gas and experience no analgesia at all, just a bit of nausea.

The problem was that there was little understanding of the mechanisms behind how ether worked, no standardisation of delivery – some doctors swore by oiled silk bags, others by contraptions with masks and valves, others a full helmet that encased the patient's head – and limited grasp of how diethyl ether could be rendered a gas and kept that way. Dr John Snow, the man best known for the discovery that cholera was transmitted by microbes in water and for his germ theory,* was one of the first to really investigate, in a scientific way, what ether did. But Snow's work was exceptional, literally – no one else was quite that methodical – and this meant the medical marketplace could be a dangerous place.

This was symptomatic of the time. Excitement about a new drug or therapy outpaced investigations into its safety or how it worked. Although, to be perfectly honest, we still don't have a great handle on how most anaesthetics work beyond a general 'disrupts pain signals in the brain and body'. The exact mechanisms of modulation and mediation have yet to be conclusively determined for a host of psychoactive drugs that can produce the effects of general anaesthesia,† but the current theory is that these different chemical compounds all act by binding with certain proteins in certain nerve cells, effectively silencing communication. In any case, not knowing didn't stop new

* And immortalised as the namesake of a mediocre Soho pub.

† And there are a lot. We've only talked about nitrous oxide, ether and chloroform, but there's also halogenated hydrocarbons, ketamine, propofol (what killed Michael Jackson), etomidate, alcohol, dextromethorphan, barbiturates… The list is very long.

drugs and formulas from popping up all the time then and it certainly doesn't now.

After the success of ether, dentists and surgeons revisited nitrous oxide as an anaesthetic, with greater success. Meanwhile, local anaesthetics, or topical pain relievers, were also coming to market and growing in popularity, including the still popular benzocaine, first synthesised in 1890 by German chemist Eduard Ritsert and sold as Anästhesin. Chloroform ($CHCl_3$), a naturally occurring gas that was less flammable than ether but more toxic, was synthesised in around 1831 and first used in minor surgery in 1847. The drug proved increasingly popular. In 1853 Queen Victoria was administered chloroform by Dr Snow for the birth of Prince Leopold, erasing any moral concerns that anyone might have had about the appropriateness of pain relief in childbirth.* If extinguishing the 'pain of Eve' was good enough for the Queen who lent her name to a period of heightened moral policing, then it was good enough for everyone else.

The value of pain was gone.

The first opioid epidemic

'Dear Sir,' the letter to the editor of the *New-York Daily Times* begins.

> I have noticed, with gratitude, your occasional efforts to awaken attention to the powerful mischief growing out of the habit of using Opium, which seems to be greatly on the increase in this country … If anybody is to be pitied beyond others, it is the user of Opium … I know of one case. A gentleman was afflicted for several years with Neuralgia. His sufferings were intense and his physician prescribed Morphine. The doses increased until they became enormously large. The disease ceased after nearly four years of torture, but the poor sufferer had contracted what was far worse, a nervous condition which demanded the continuance of the Opium. He deeply

---

* Despite many claims to the contrary, Queen Victoria did not use cannabis to relieve her menstrual cramps or manage pain post-delivery; she didn't smoke it, she didn't eat it, she didn't take it in a 'tincture'. Victoria was not on the pot.

> lamented his bondage and made desperate efforts to free himself, but in vain. I suppose no one can imagine the distress which the denial of the drug occasions except those who have tried it.

The letter was signed 'Yours, Candor.'

It's quite possible that Candor was the 'gentleman' in question, given how sympathetically he or she writes about the plight of the opium addict. But many people were touched by opium addiction, whether it was themselves or a loved one. In 1852, just as it is now, America was in the grips of an epidemic* of opioid addiction, driven by pain and over-prescription.

At the time Candor wrote, the *New-York Daily Times* – or the *New York Times*, as it would soon be known – was concerned about opium, but they were among the few. Opium, in 1852, was perfectly legal, fairly cheap and widely available in chemists, grocers and pharmacies in the US. In an editorial that appeared on the front page in August 1852 – quite possibly the editorial that prompted Candor's letter – the paper expressed shock that some 'medical writers' had concluded that opium consumption was 'less injurious to social morals than alcoholic drinks'. This was on the wisdom that you didn't tend to see opium eaters starting fistfights outside bars, hitting their wives or stealing. Though it might have been true that alcohol was more detrimental to the social order, the paper agreed, opium wreaked havoc on the 'general health of the individual' and their 'intellectual power' and was much harder to quit. The paper appealed to the state government to limit or regulate the sale of opium and in the meantime, given that that might prove a lengthy process, to 'friends of humanity'. But as worrying as opium was starting to become, sales of it had also made a

* Let's talk briefly about the word 'epidemic'. Epidemic, scientifically speaking, refers to the usually sudden outbreak of an infectious disease in numbers greater than would be expected in that area. In that regard, using a phrase like 'opioid epidemic' is technically a misnomer. However, it is sticky, not least because of the near-alliteration and the fact that it sounds super-scary in a way that 'public health emergency' rather doesn't. Both the Centers for Disease Control and Prevention and the US Department of Health and Human Services – two organisations that should certainly know the definition of epidemic – refer to the current crisis as an 'epidemic' and so does everyone else. We'll go with epidemic.

lot of people rich; that alone tended to put the brakes on efforts to reform or regulate the market.

Opium and its derivatives come from the latex residue of the unripened seed-pod of the poppy plant or, in Latin, *Papaver somniferum*. Though selling it has changed, harvesting opium has not in millennia: workers score a shallow cut in the seed-pod, allowing the milky sap to leak out and dry. Once it's turned brown and gummy, the resin is scraped off with a carved wooden spatula and allowed to dry further in open boxes. The resulting resin is rolled into balls and can be sold in this raw form. Raw opium, though bitter, can be drunk, swallowed or smoked, and it induces a feeling of relaxation, mild euphoria and enhanced mood – the ancient Sumerians didn't call it the 'joy plant' for nothing. It was also popular with manual labourers because it suppressed the appetite while stimulating activity, making it possible to work longer on less fuel. And, of course, it works as an analgesic.

The ancient Egyptians pioneered the opium trade, introducing it to Greece, northern Africa and Europe; while Europe stumbled through the 'Dark Ages', Arabic medicine made liberal use of opium, ensuring the practice would spread throughout the East. By the time Elizabeth I was launching her ships in the direction of new trade and the new world, opium was well established across the Middle East, India and Asia, and European markets wanted it. And by the 19th century, world powers were willing to fight for it: in 1839, Britain went to war with China to protect its opium trade. China had banned the opium trade, but the British disagreed; when the Ch'ing Dynasty seized British opium stocks in Canton, the British responded with cannons. The first war ended in a British victory in 1842 but it picked up again in 1856. The second opium war ended in 1860 with the legalisation of opium imports.

Globalisation exploded the opium trade in much the same way it did alcohol and tea. But though some of opium's use was recreational, much of it was medicinal, used to treat chronic or acute pain conditions, anxiety, intestinal discomfort. Use of the drug was both on the advice of physicians and on the advice of popular wisdom; with no restrictions on who could obtain it, people diagnosed and treated themselves (and all *without* the benefit of the internet!). Laudanum, a heady mixture of 10 per cent opium and 90 per cent alcohol first invented in the

16th century, was devastatingly popular. Its increased availability in the 19th century meant that 'nervous' people and insomniacs drank it as an evening digestif, nurses dosed 'troublesome' infants with it, and women regularly used it to manage menstrual cramps. It was used as a cough suppressant, often providing the only relief for people dying of tuberculosis. Though opium couldn't cure any underlying pathology, it was effective in mitigating the symptoms of a variety of other maladies, too. One of the side effects of opiate use is constipation, making opium a treatment for diarrhoea; it was considered a useful treatment for malaria.

Though people taking the drug in the 18th and 19th centuries didn't know it, opium stifles pain by mimicking the effects of our endogenous opioids. Poppy resin contains several alkaloids, plant-made nitrogen-based compounds, some of which possess pharmacological qualities; caffeine and cocaine are both alkaloids. The most potent of opium's alkaloids is morphine, which makes up 10 to 16 per cent of the opium resin. We humans come equipped with at least five different types of receptors that respond to opioids, including mu, kappa, delta, nociception, and zeta receptors; morphine works in part by binding to mu receptors and disrupting nociceptive signalling in both directions. This means that afferent signals from the periphery – the alarm bell that something nociceptive is happening – don't make it to the brain, and the descending signals meant to incite the inflammatory response at the site of damage are also blocked. Tissue could be damaged, or even still being damaged, but the nociceptors are not getting their message through. Codeine, another of opium's alkaloids, works in a similar fashion, binding to mu receptors; codeine is less potent than morphine but it still packs a punch. Morphine and codeine are both opiates, the classical term to describe drugs derived from opium; opioid describes any drug, naturally occurring or synthesised, that binds to opioid receptors in the human body.

However, opioids and opiates don't work by just replacing our endogenous opioids. In a study funded by the National Institute on Drug Abuse, part of the National Institutes of Health (NIH), a team led by Dr Mark von Zastrow of University of California, San Francisco found that synthetic opioids act quite a bit faster and more powerfully than our home-grown opioids. The team used a tiny sensor called a nanobody that emits a fluorescent light when an opioid receptor is activated to track activation in both endogenous opioid and administered opioid activate. This tiny sensor revealed that, while our endogenous opioids activated receptors on the cell membrane and inside the endsomes (the membrane-bound vesicles at the end of axons), opioid drugs went even further, rapidly activating receptors in cell structures called the Golgi apparatus and Golgi outposts. Von Zastrow suggests this additional activation and the speed of the response could be behind the negative consequences of opioids, namely tolerance and dependence. Adding to this, opiates and opioids also increase the circulation of everyone's favourite neurotransmitter, dopamine, which (among other things) drives the feeling of *wanting* more and more.

Raw opium has some of these effects, but scientists in the early part of the 19th century were interested in extracting the pharmacologically active components in opium. Morphine was the first opiate to be isolated: in 1803, pharmaceutical chemist Friedrich Wilhelm Adam Serturner, then just 20 years old, extracted morphine crystals from the sticky poppy juice. Over the next two years, Serturner conducted a series of experiments, dosing stray dogs and captured rats with what he was calling 'Morphium', after Morpheus, the Greek god of sleep and dreams (and not Laurence Fishburne's character in *The Matrix*). In the grand tradition of scientists experimenting on themselves, Sertürner took morphine during a terrible toothache and found himself painless and confident that morphine was safe for human consumption. Codeine followed in 1832 and soon found use as a cough suppressant (it appears to decrease activity in the part of the brain associated with the coughing mechanism). But through the 19th century, as opium and opiates became more available, a pattern was developing: more or less the route to addiction that Candor describes in his letter to the *Times*. As the paper succinctly put it in an article in 1875, 'Taken, perhaps, in the first instance to alleviate the torments of neuralgia or toothache, what proves to be a remedy soon becomes a source of gratification, which the wretchedness that follows on abstinence renders increasingly difficult to lay aside.'

People wanted and needed pain treatment. But though *overuse* clearly made addicts, no one had a good handle on what *appropriate* use looked like. Medical science was only just discovering controlled trials, but most people – pharmacists and physicians too – relied on guesswork and experience. Perhaps unsurprisingly, tragic stories of overdoses, both accidental and intentional, frequently appeared on police report blotters. In one such report from 1851, a woman gave her child a dose of opium to calm him, 'without knowing its effect', then gave herself some to help her sleep. The child was found dead and the woman was only barely revived.

Other technological advancements turned opiates into a different kind of problem. The invention of the hypodermic needle, attributed to Scottish surgeon Alexander Wood, changed the delivery of the drug and rendered it more potent. The syringe was introduced to America in 1856, just in time for the Civil War. During the war, the Union Army alone handed out nearly 10 million opium pills to its soldiers, as well as 2.8 million ounces of opium powders and tinctures, according to David Courtwright, historian and author of *Dark Paradise: A History of Opiate Addiction in America*. 'Even if a disabled soldier survived the war without becoming addicted, there was a good chance he would later meet up with a hypodermic-wielding physician,' Courtright wrote. 'Though [morphine] could cure little, it could relieve anything … Doctors and patients alike were tempted to overuse.'

At the same time, it wasn't just opiates: other drugs were appearing in the medical – and recreational – marketplace. Just as was the case with anaesthetics, chemical discoveries prompted a lot of excitement but less caution. Cocaine came into medical use as a local anaesthetic, cough suppressant and general 'invigorator' in the 1880s, after Albert Friedrich Emil Niemann, a German chemist, and Paolo Mantegazza, an Italian neurologist, isolated the cocaine alkaloid from the coca leaf around the same time in the late 1850s. Niemann published his findings in his dissertation in 1860, noting that the white powder produced numbness and a sense of cold when he placed it on his tongue. Today, we know that cocaine is part of the amino ester group and works as a local anaesthetic by blocking sodium channels, so pain signals never reach our awareness. Cocaine also blocks the reuptake of dopamine into the transmitting neuron, meaning more remains in the synapse. This build-up of dopamine in the synapse is what can produce a sense of pleasant alertness and wanting for more.

Cocaine proved particularly practical in surgery: ether and chloroform were useful anaesthetics during many kinds of surgery, but their tendency to induce vomiting in patients made them less than ideal for more precise operations, such as eye surgery. In 1884, Karl Koller, a pioneering ophthalmologist, soaked his patient's eyeball with a solution of cocaine; the effect was immediately anaesthetic, ushering in a new era of painless minor surgeries. But cocaine was also massively popular in less salubrious ways, used in everything from wine to hair tonic, and in combination with other drugs. In 1908, the *New York Times* reported that the drug became even more popular once people discovered that the 'terrible reactions which are suffered by the morphine and opium taker could be entirely overcome by a little cocaine.'

Cocaine, opium and opiates, and alcohol were also frequently used in proprietary 'patent'* medicines, compounds of unknown ingredients that promised miraculous cures for a wide variety of illnesses. These were hugely popular and had names like Daffy's Elixir Salutis, Dr Bateman's Pectoral Drops, John Hooper's Female Pills, Lydia Pinkham's Vegetable Compound, Hamlin's Wizard Oil, and Dr Newman's Famous Anti-Venereal Pill. Most were ineffective, some were lethal, all were unregulated – drugs, medical devices and 'cures' were being sold in a marketplace devoid of nearly any oversight. That consumers had access to many potentially deadly, certainly addictive drugs and little idea of how to use them safely was one problem; that no one really knew what was in John Hooper's Female Pills or Hamlin's Wizard Oil was another; and that they couldn't always be certain it *was* morphine or cocaine that they were buying was still another.

Something needed to be done, but exactly what was the subject of fierce debate. Efforts to standardise drugs and their preparations were earnest but sporadic and reflected the vying for territory between

* The word 'patent' here doesn't actually mean approval or licence granted by any authority – chemical patents didn't even exist in the US until 1925 – but rather it refers to 'patents of royal favour'. These were claims, usually made in bold letters on the bottle, that the medicine was made by those who had a royal warrant to provide medicine to the royal family. So basically, a celebrity endorsement.

physicians and pretty much everyone else involved in the expanding health and wellness landscape. As physicians both in North America and Europe attempted to consolidate their power and control over all things to do with the body, chemists and pharmacists also tried to carve out their patch. The Royal Pharmaceutical Society was founded in 1841 to introduce qualifications for chemists and pharmacists and, to some degree, to draw a distinction between them and physicians. The Adulteration of Food and Drugs Act 1860 had established the office of Public Analyst, to ensure the quality of food and drugs, but that didn't affect the sales of drugs. In 1868, the UK passed the Pharmacy Act, which decreed that opium and other drugs could only be sold by licensed pharmacists, but this legislation was only lightly enforced until 1908, when harsher penalties were introduced.

In the US, similar efforts to restrict sales of drugs were comparably toothless. Drugs were big business: between 1870 and 1890, imports of medicinal opiates to the US doubled. Importation of raw opium to the UK stood at 114,000 pounds in 1852 but rose to 356,000 pounds in 1872, according to an 1875 *New York Times* article. When in 1887 the US Treasury discovered a trove of 1,032 opium samples, weighing 2.5 ounces each, that had been collected for taxation purposes but never destroyed, they decided to sell them for a 'four-figure' windfall, the *New York Times* reported.

Opiates and cocaine, like alcohol and gambling, were both morally questionable and commercially encouraged, both socially deplored and socially endorsed; unlike alcohol and gambling, however, drugs had a medicinal value. *Who* was an addict was increasingly dressed in moral rhetoric (just as it is now). Middle- and upper-class women who were opium addicts were guilty of a bad but genteel habit. People in lower classes, however, were 'abusing' the drug. What kind of addict you were also depended on your race, according to the racist, eugenics-driven wisdom of the time: where people in China and Malaysia who were addicted to opium were 'pitiable creatures', those of Anglo-Saxon stock were 'pleasant and conversable', according to an 1875 *New York Times* editorial, though liable to fall into a deep and miserable depression after the drug wore off.

With regulating the drugs market a contentious business, some pharmaceutical entrepreneurs decided to throw 'good' drugs after bad. Diamorphine, also known as morphine diacetate or diacetylmorphine, was first synthesised in 1874 but it wasn't until 1895 that the drug made it onto the market – as 'heroin', marketed explicitly as a non-addictive alternative to morphine. Of course, it was a far more potent and far more addictive opioid – the addition of the acetyl functional groups increases the solubility of the now acetylated morphine molecule in fats. This helps it cross the blood–brain barrier more quickly; once on the other side, the acetyl groups fall away and the morphine is left to do what it does and bind to opioid receptors. People reached that euphoric state much more quickly than on morphine alone; heroin married to hypodermic drug delivery became an incredibly potent combination. The problem with this became disastrously apparent within a few years, yet the drug was still considered a safer alternative to morphine into 1910.

That heroin – *heroin* – could plausibly be marketed as non-addictive and better than morphine captures some of the chaos of the pharmaceutical markets at the turn of the century. Regulation was amping up but lagged behind the rapid-fire chemical inventions pumped out by new pharmaceutical companies. People still needed and wanted drugs – by now, pharmacological relief was an expectation they were willing to pay for – but drug makers weren't, at this point, required to demonstrate that their drugs were safe or even effective. This new era of drug invention and distribution was drawing novel intersecting lines around the chemist, the pharmaceutical company, the doctor, the scientist and the patient in the arena of pain management, and who was in charge of doling out this pharmacopeial revolution was still in question.

Eventually, concern over dangers of unknown chemical compounds and the addictive properties of drugs spurred the passing of the 1902 Biologics Control Act in the US, the first of many legislative acts aimed at controlling the things we put on our bodies and in our mouths. Four years later, President Theodore Roosevelt signed into law the Pure Food and Drug Act, requiring the labelling of any medicine or food that contained cocaine, opium, opiates, cannabis or alcohol, and laying the foundation for what would become the Food and Drug Administration.

And, amid what had become a national panic over the 'drugs craze', in 1914 the Harrison Act banned unlicensed sales of opiates and cocaine, effectively criminalising people who used or were addicted to the drugs. In the 1920s the UK passed legislation regulating the possession and use of 'dangerous drugs', including cocaine, cannabis and opiates, although doctors retained the right to prescribe essentially anything they wanted until the 1960s.

The promise of a pain-free life

Ultimately, however, cocaine and opium's withdrawal from the legal market wasn't only down to moral outrage, the expansion of governmental powers or the sympathy of the concerned anti-addiction movement. With improving sanitation and a wider understanding of germ theory, some illnesses that would have been met with opiates were in retreat. Meanwhile, science-driven medicine was rapidly finding causes and cures for a number of diseases, diminishing the need for these drugs. As Courtwright wrote in *Dark Paradise*, 'The achievement of greater diagnostic precision, made possible by the discovery and classification of pathogenic microorganisms by the development of new techniques, such as x-radiation, discouraged the unthinking palliation of disease; doctors who shot first and asked questions later were increasingly criticised for masking the symptoms of illnesses otherwise diagnosable and treatable.' More and more, would-be doctors were cautioned against the overuse of morphine; the best doctors were the ones who used opiates as little as possible.

At the same time, research was uncovering new, non-narcotic and less dangerous analgesics. And with plausible alternatives, albeit some with less euphoria, anti-opium activists could make the case that opium was no longer necessary. Widely available over-the-counter analgesics were one of the great inventions of the 20th century, a nearly unrivalled success story; their development assuredly saved some people from opiate addiction. But like opioids, over-the-counter drugs weren't without their dangers.

One of the first non-opioid pain relievers brought to market was phenacetin, an aniline derivative with fever-reducing properties marketed by Bayer in 1887. The drug was widely used through the 1940s and 1950s; evidence suggests it worked as a pain reliever by increasing the concentration of amine transmitters, such as serotonin, in the brain and spinal cord, which modulates the incoming pain information. But chronic use had a tendency to cause severe renal damage – billionaire-turned-recluse Howard Hughes's death from kidney failure may have been caused by his use of phenacetin to manage his chronic pain – and was linked to renal and ureter cancers. The drug was banned from the Canadian market in 1973 and from the US 10 years later.

Luckily, there was another aniline derivative ready to step in – paracetamol, also known as acetaminophen. Paracetamol was actually first synthesised in 1877 but was dismissed after one study indicated it induced a serious blood disorder, methemoglobinemia, in which red blood cells aren't able to release enough oxygen to cells. It didn't, but this wasn't known until 1948, when scientists also discovered that phenacetin worked as an analgesic because it metabolises in the body into paracetamol. Paracetamol eventually made it to market in combination with aspirin and caffeine as Triagesic in 1950 in the US, but it was pulled the following year when three people who used it came down with what turned out to be an unrelated blood disease. In 1955 McNeil Laboratories introduced it under the name Children's Tylenol Elixir and concerns over the safety of the drug began to wane. By the 1980s, paracetamol was a popular fever reducer, outselling aspirin in the UK.

And that's saying something because, globally, aspirin is the biggest over-the-counter pain reliever in the world. Willow bark, high in the alcoholic glucoside salicin, had been used medicinally for millennia as an anti-inflammatory and fever reducer. Throughout the 19th century, chemists worked on isolating the active ingredient from the plant; German chemist Johann Andreas Buchner was able to extract salicin from willow in 1828. Salicin can be chemically oxidised to make the enormously useful salicylic acid, and in 1859 another German chemist, Hermann Kolbe, building on the work of other scientists, successfully synthesised pure salicylic acid. This synthesised version, however, caused horrible stomach irritation and vomiting. Salicylic acid wasn't exactly put on the shelf – it was used as an antiseptic in surgeries, in treating skin conditions and in preserving food – but it took another 40 years before Kolbe's work would result in aspirin.

In the 1890s, working at the direction of Arthur Eichengrün, Felix Hoffman (the same Bayer chemist who created heroin) added an acetyl group to the salicylic acid, reducing the unpalatable side effects. Bayer quickly patented the new process and Aspirin – now a trademarked name – was on the market by 1899; within 50 years it became, according to Bayer, the most frequently sold painkiller in the world. It wasn't until the 1970s, however, that anyone really figured out *how* aspirin worked to inhibit pain: John Vane, professor of pharmacology at the University of London, shared the Nobel Prize in Medicine in 1982 for his discovery that the drug inhibits the body's cyclooxygenase enzymes, responsible for the synthesis of the hormone-like lipid compound prostaglandin, which can cause inflammation and fever.

Unfortunately for aspirin users, this action is also what causes the drug's negative side effects: prostaglandins are also responsible for protecting our stomachs and maintaining kidney function. Ibuprofen (or isobutylphenyl propionic acid), which was developed in the 1960s by Boots pharmaceutical researchers as an alternative to aspirin, works in a similar fashion. And though its gastrointestinal side effects are meant to be less severe, ibuprofen can still cause them.

In fact, all NSAIDs and over-the-counter drugs come with side effects, some of them severe; this was the case throughout the 20th century and it's the case now. Chronic use of NSAIDs increases the risk of developing peptic ulcer disease, experiencing acute renal failure, and stroke or myocardial infarction; it can also exacerbate a number of chronic conditions, including heart failure and hypertension. This is particularly dangerous for the population of people who rely on NSAIDs the most: older adults. Individuals over the age of 65 are at a greater risk of adverse drug reactions, owing to age-related changes in pharmacokinetics – an exciting way of saying how the drug moves through and is processed by the body – comorbid conditions, and any other drugs they might also be taking. Studies have demonstrated that NSAID use increases the risk of fatal peptic ulcers in older adults by three to five times; overall, it's estimated that NSAID use causes 41,000 hospitalisations and 3,300 deaths each year among older adults.

We are probably making this worse: evidence suggests we are taking more than we should. A 2018 Boston University survey of

1,326 ibuprofen users found that not only did close to 90 per cent of those surveyed take ibuprofen in the week of the survey, but also 15 per cent of the respondents admitted to taking more than the recommended dose of ibuprofen and other NSAIDs. And it's not just the US: France saw a 53 per cent increase in the use of paracetamol between 2006 and 2015, with consumers also taking higher doses – consumption of the 1000mg tablets grew by 140 per cent, as compared to only 20 per cent of the 500mg tablet packs. This is more dangerous than we acknowledge: paracetamol overdose is one of the leading causes of liver failure, but this is complicated by the fact that the distance between a 'safe' dose and a 'toxic' one is narrow and variable. According to the British Liver Trust, even taking one or two pills more than the recommended dose can cause serious liver damage.

Perhaps most frustrating is that all these drugs might not be giving us the relief we're seeking. Recently, a Cochrane review, an internationally recognised systematic overview of healthcare research, found that paracetamol for lower back pain worked only as well as a placebo treatment. Another fun fact: a 2017 University of Greenwich study found that sinking two pints of beer was 24 per cent more effective at providing pain relief across multiple modalities – thermal pain and pressure pain – than paracetamol.

The bottom line is that over-the-counter drugs – when they work and we use them correctly – are safer than opioids and less addictive. But that doesn't mean we should keep using them the way we do. In an NBC News report on the study that found Americans are using too much NSAIDs, some doctors suggested NSAIDs shouldn't be available over the counter because consumers simply don't know how to use them properly.

That said, a big part of the reason NSAIDs are so popular is because many people are living with a lot of chronic pain and very few options to manage it effectively. 'When narcotics are not being used to manage such pain, NSAIDs often are,' Dr David Katz, who heads the Yale University Prevent Research Clinic, wrote in an email to NBC News. 'That a substantial subset of those relying on NSAIDs are using them ill-advisedly or excessively is rather to be expected under these circumstances.'

Binge and bust

NSAIDs came to dominate the medical marketplace in the 20th century but they were part of a bigger shift in the practice of medicine. Despite the recognition that all drugs came with consequences and side effects, demand for pharmacological fixes was high. This was something concrete that doctors could offer to patients in pain, that played well in a biomedical model of the body and health, and that people *wanted*. Scientific advances came at a breakneck speed – between 1938, the year it was created, and 1962, the FDA fielded 13,000 new drug applications, for everything from antibiotics and pain relievers to psychotropics and tranquilisers. Medical historians believe that about 90 per cent of the drugs prescribed in 1960 had been introduced in the previous 20 years; 40 per cent of them could not even have been filled in 1954.

This development was watched with concern by some – newspaper headlines worried that Americans were on an 'aspirin binge' after it was revealed in 1960 that consumption of the pain-relieving drugs had increased four times as fast as the population growth, while the *Saturday Review* described the US as 'the most over-medicated country in the world' in 1962. But the fact was that pharmaceuticals, especially prescription drugs, were a huge market, far bigger than patent medicines and opiates and hair tonic had been. Sales of tranquilisers in the US grew from $4 million in 1960 to $8 million in 1964. Between 1939 and 1959, drug spending increased from $300 million to $2.3 billion, and all but about $4 million of that increase was due to prescription drugs.

Once the juggernaut of pharmaceutical profit and pain relief got rolling, it was difficult to put the brakes on and political will to do so waned and waxed depending on who was in office. In 1938 President Franklin Delano Roosevelt signed the Food, Drug and Cosmetic Act into law, creating the Food and Drug Administration (FDA) and igniting what is still an ongoing, often contentious power struggle between doctors, pharmaceutical companies, health insurance companies, government officials and consumers. The Act increased federal oversight, specifically requiring drugs to go through a pre-market safety review, and banning the use of false therapeutic claims regardless of whether there was fraudulent intent. The 1951 Durham-Humphrey Amendment attempted to clarify the drug landscape by dividing it into over-the-counter and prescription, which were considered unsafe for self-medication, habit-forming or otherwise potentially harmful. The Kefauver-Harris Amendment in 1962 required all new drug applications to demonstrate 'substantial evidence' of efficacy for the marketed indication. And in 1969 the FDA announced that advertisements must offer 'true statement of information in brief summary relating to side effects, contraindications, and effectiveness.'

But the willingness to regulate the market depended on the political landscape. In 1980s America, pharmaceutical companies found a friendly environment under the more hands-off Reagan administration, as well as a few loopholes in FDA advertising requirements. Against the backdrop of the patient-to-consumer shift we talked about in the last chapter, the first pharmaceutical adverts began appearing on TVs, in print and on radio in the 1980s – thinly disguised as public service adverts informing people about health conditions. Health conditions that drug companies just happened to have drugs to treat. Pfizer, for example, ran campaigns about hypertension and diabetes, urging people to 'talk to their doctors'. Who could argue against encouraging people to take ownership of their health? FDA commissioner Arthur Hull Hayes, for one. Recognising that pharmaceutical companies were just trying to advertise directly to the patient, he argued in 1986, '[T]his form of advertising could lead patients to pressure physicians to prescribe unnecessary or unindicated drugs, increase the price of drugs, since consumer advertising is generally much more costly than physician advertising, confuse patients by leading them to believe that some minor difference represents a major therapeutic advance, potentiate the use of brand name products rather than cheaper, but equivalent, generic drugs, and foster increased drug taking in an already overmedicated society.' Doctors protested, too; a letter published in the *New England Journal of Medicine* in 1982 worried that advertising directly to consumers 'may tend to undermine physician control over prescribing' and that 'most lay people are ill equipped to evaluate the efficacy or toxicity of drugs.'

The thing was, despite the language of empowerment that accompanied the patient-as-consumer paradigm, patients didn't feel comfortable with the new adverts either. A survey conducted in the 1980s reported that more than 75 per cent of consumers didn't think they could determine whether they should use a specific drug based on an advert, and 63 per cent didn't think they could tell if an advert was misleading. Patients, now speaking as consumers, wanted – and believed, correctly, they had a right to – more information about the drugs they were being prescribed. But they were also trying to orient themselves in a medical landscape that now included the health product market. This category included everything from skin care to smoking cessation aids, cough medicine to toothpaste, diet drugs to vitamins, and it was swelling year on year to serve this new consumer. According to the Consumer Healthcare Products Association, American spending on over-the-counter products, including things like itch creams, pain relief and acne medications, was $5.5 billion in 1980. By 1996 that figure had risen to $16.5 billion.*

Nevertheless, in 1997 the FDA published new guidance on drugs advertising that effectively relaxed regulations. Now all pharmaceutical companies had to do was introduce the symptoms and the prescription drug, invite the viewer to ask their doctor about it, and then run through a list of a few side effects at an auctioneer's speed. Spending on television drugs adverts increased from $310 million in 1997 to $667 million in 1998, and by 2005 total spending was $3.3 billion, up from $1.3 billion in 1998. And it was incredibly effective: in 2002, 98 per cent of consumers surveyed reported they had seen or heard an ad for a prescription drug. Of those who had seen an ad, an astonishing 33 per cent *did* ask their doctor about the drug and 79 per cent of those left with a prescription for it.

By far the biggest success story of the growth of the pharmaceutical industry was in, you guessed it, opioids. Regulation in the early part of the 20th century tried to end self-medication with opiates, but

* Also according to the Consumer Healthcare Products Association, Americans spent $41 million on enema products in 2018, about $1 million more than on ear drops. There's got to be a joke in here somewhere but we're not certain we're qualified to make it.

pharmaceutical entrepreneurs were by no means willing to give up on developing ever more powerful opioid-based medications. Over the course of the century, drugs makers developed oxycodone (1916), hydrocodone (1920), hydromorphone (1924), meperidine (1932), fentanyl (1959) and etorphine (1960), among many others. To put in perspective just how powerful these drugs are, etorphine is 1,000 to 3,000 times more powerful than morphine and is only legally used to knock out elephants.

However, prior to the mid 1990s, powerful opioids were used only in the most extreme circumstances, such as late-stage cancer. This limited use changed when pharmaceutical companies hijacked a one-paragraph letter to the editor published in the *New England Journal of Medicine* in January 1980, which reported only 1 per cent of patients prescribed narcotics became addicted. Purdue, the maker of OxyContin (the brand name of oxycodone, which is twice as powerful as morphine) used this anecdotal, inaccurate observation to spearhead one of the most successful marketing campaigns ever seen. The evidence of their success: the crippling opioid epidemic.

It wasn't *all* Purdue, but the company was perhaps the biggest influence in the opioid market. Purdue's strategy for OxyContin was to target advertising to doctors and consumers. Between 1996 and 2001, Purdue hosted more than 40 all-expenses-paid 'pain management' conferences for US physicians at resorts in Florida, Arizona and California. Between 1996 and 2001, their sales team swelled from 318 to 671 and their call list went from 35,000 physicians to almost 100,000. Sales reps systematically went through prescribing data zip code by zip code, finding the highest and lowest prescribers, building profiles on them, and targeting their practices, handing out 'patient starter' coupons good for a seven- to 30-day supply of OxyContin. OxyContin was being used to relieve pain from knee surgeries, menstrual cramps and back aches; from 1997 to 2002 prescriptions for oxycodone increased 402 per cent. But it's not that OxyContin was *that good* or even any better than any other opioid. The drug's biggest asset was its slow-release delivery; a medical reviewer officer for the FDA concluded in 1995 that it had no significant advantage other than reducing the number of daily doses, and a 2001 report confirmed OxyContin carried no higher therapeutic effect compared to other available opioids.

Maine, West Virginia, eastern Kentucky, southwestern Virginia, and Alabama were heavily targeted, and it worked; these areas had prescribing rates between two and five times the national average. The consequences were swift: Maine saw a 460 per cent increase in the number of patients treated for opioid abuse between 1995 and 2001; during that same period, eastern Kentucky saw a 500 per cent increase in the number of patients entering methadone programmes – 75 per cent of those were OxyContin-dependent. Between 1997 and 2003, deaths from drug overdose in rural Virginia increased 300 per cent; prescription opioids accounted for the vast majority, at 74 per cent of those deaths. By the beginning of the 21st century, OxyContin had a new name: 'hillbilly heroin'. By 2004, a year in which Purdue spent $200 million in promotion, OxyContin was the most abused drug in the United States.

A high tide raises all boats and it wasn't just sales of OxyContin that went up. From the mid-1990s, fentanyl and morphine were also being prescribed at astonishing rates; the number of prescriptions issued for opioids such as Vicodin and Percocet skyrocketed from 76 million in 1991 to nearly 207 million in 2013. The results were predictable and depressing: the Drug Abuse Warning Network found a 641 per cent increase in emergency room visits for fentanyl alone in this period.

It's important to note that some research indicates that many people who became addicted to OxyContin didn't begin using the drug to treat or manage pain. A 2007 University of Pennsylvania study published in the *American Journal of Psychiatry* evaluated OxyContin use in 27,816 people admitted to treatment programmes between 2001 and 2004; it found that only 22 per cent of them said they got the drug through a prescription. Now, here's the big grain of salt: Purdue, in the midst of a lawsuit over their marketing of the drug, paid for that study. The company was attempting to make the argument that they weren't responsible for turning people into addicts, rather addicts had stolen their drug and were now ruining *their* reputation and the reputations of people in real pain.

But even if people were getting OxyContin from sources other than legit doctors – from friends and family with prescriptions, from unethical doctors running 'pillmills', doling out prescriptions indiscriminately for pay, or organised crime – the increased availability of the drug in general made that possible. The Department of Justice didn't buy Purdue's defence and in May 2007, Purdue Frederick Company Inc., an affiliate of Purdue Pharma, and three Purdue executives pleaded guilty to criminal charges of misbranding OxyContin by claiming it was less addictive. Their sentence was $634 million in fines. Which sounds like a lot. But between 1996 and 2006, Purdue Pharma made between $1 to 2 billion a year on the sale of OxyContin; their fine averaged between 5 and 8 per cent of their profits. The company easily recouped its losses and then some, earning $2.8 billion in 2011 alone.

However, the opioid-related lawsuits kept coming. Facing literally thousands brought by states, cities and counties, Purdue quietly announced it was filing for bankruptcy on 15 September 2019. This, the company determined, was the best decision 'for the people'. Steve Miller, the chairman of the company's board of directors, said in a statement, 'This settlement framework avoids wasting hundreds of millions of dollars and years on protracted litigation, and instead will provide billions of dollars and critical resources to communities across the country trying to cope with the opioid crisis.' Which is, in the opinion of the many people whose lives and loved ones were destroyed by OxyContin, utter horseshit. In February 2020, negotiators for those hurt by Purdue conceded that the settlement they were looking at was likely to be far less than they'd hoped.

Efforts to hold Purdue responsible might be a case of too little, too late. The US is now the biggest global consumer of opioids, accounting for nearly all of the world's prescriptions for hydrocodone and 81 per cent of oxycodone (Germany is the second-largest consumer but Germans seem to prefer fentanyl, which is cheaper and can be administered via patch). Between 2016 and 2017 in the US, more than 130 people died each day from opioid-related drug overdoses, according to the US Department of Health and Human Services. The White House Council of Economic Advisers puts the total cost of the premature opioid-related deaths from 2015 to 2018 at more than $2.5 trillion. No settlement can mitigate that.

It's hard not to look at the US government's efforts to regulate and control the production, sale and use of drugs and think that the health and welfare of its citizens is lower down the list of concerns than perhaps it should be. But this is not just a US problem. In 2018 the UK's NHS came under fire from addiction specialists who accused the service of over-prescribing opioids and 'creating drug addicts', according to a BBC report. While the Royal College of GPs denied doctors were prescribing opioids as a 'quick fix', nevertheless, opioid prescriptions in the UK increased by 34 per cent between 1998 and 2016. That might not sound like a lot, but when the figure is adjusted to reflect total oral morphine equivalency, that increase was 127 per cent – basically, doctors are prescribing stronger and stronger drugs. One doctor told the BBC that GPs are increasingly 'under pressure and patients demand they get given these strong painkillers.' In 2019 the *Telegraph* reported that the number of prescriptions written for opioids in the UK had risen from 89 million in 2008 to 141 million in 2018, a 60 per cent increase the paper attributed to 'middle-aged women'. (If you're hearing echoes of the genteel 19th-century addiction to laudanum that seemed to affect mostly women, well, you're on the right track.)

Meanwhile, as doctors in the US have started to more tightly control opioid prescriptions, drug makers are marketing them more aggressively elsewhere, especially in Europe and Latin America. A 2017 study analysing surveys conducted in high- and middle-income countries in Europe, Latin America, the Middle East and Asia, found that opioid abuse rates among youth are growing substantially. One of the biggest problems for pain treatment in developing nations is that pain-relieving drugs, already in short supply, are diverted from the people who need them and funnelled into the hands of people who don't.

Frankly, addiction isn't the only problem: fear of abuse has led to inadequate access to pain relief, as Jude Taylor found when she ended up in the emergency department suffering from endometriosis. 'The still-growing US opioid epidemic lies at the intersection of two substantial public health challenges: reducing the burden of suffering from pain and containing the rising toll of harms associated with the use of opioid medications,' a team of experts from the National Academies of Sciences, Engineering, and Medicine noted in the lead article in a highly influential 2019 special edition of the *American Journal of Public Health*. The edition was intended as 'a platform for commenting on the public health burden of pain', but in addition to outlining the reams of research demonstrating that opioids are a poor choice to address chronic pain, they also offered a sharp critique of the Centers for Disease Control's updated opioid guidelines. The new guidelines, published in 2016, prioritise curtailing opioid prescriptions, as a way to address opioid use disorder, over treating chronic pain. While both initiatives are important, there is real concern that the guidelines 'are being adopted arbitrarily by regulators and healthcare organizations to reduce opioids unilaterally without addressing the need for expanded resources of integrated pain care,' the authors wrote. 'Such unilateral restrictions of opioids may be driving additional people into illicit opioid markets with even greater harm.' In other words, when the pendulum swings, it sometimes swings too far.

The impossible dream?

Opioids get a lot of attention – and rightly so – but they're not the only example of the need to treat pain (especially chronic pain) outstripping caution. One promising direction stemmed from understanding chronic neuropathic pain as the result of increased sensitisation in the nervous system or, to put it another way, hyperexcitable neuronal activity. Maybe, researchers thought, treatments shown to be effective in treating epilepsy, another overexcited brain condition, could be helpful to manage neuropathic pain? Gabapentin, an anticonvulsant, has been available for epilepsy since 1993; it was the first in a class of drugs now called gabapentinoids, which inhibit specific calcium ion channels. It seemed to work – in 2002, after several trials, gabapentin was approved in the US for use in treating neuropathic pain resulting from reactivation of the varicella zoster virus, also known as shingles.* Pregabalin, meant to be gabapentin's successor, has been available since 2004. Gabapentinoids were ultimately approved to treat four kinds of neuropathic pain – shingles-related neuralgia, diabetic neuropathy, fibromyalgia and spinal cord injury.

Finally, it seemed, there were treatment options that didn't rely on addictive opioids and seemed to target the root of the problem, even

* Shingles sucks.

legitimising conditions that have been chronically disbelieved. In June 2007 the FDA green-lit Lyrica, Pfizer's pregabalin-based anticonvulsant, for treatment for fibromyalgia pain, the first-ever drug specifically approved for the illness. This was, in a word, controversial (not least because pregabalin was at one point the second-most expensive drug available on the NHS). Pfizer had been aggressively promoting gabapentin, under the name Neurotonin, and later, Lyrica, for uses well off-label *before* the drugs were approved to treat a few neuropathic conditions. Prescribing drugs for conditions they have not been approved to treat is legal and very common; drug licences and approvals can be narrow, not to mention expensive to achieve, so it's not surprising they don't always encompass the breadth of everything a specific drug can treat. This, however, puts the onus on the prescribing doctor – who is just as susceptible to marketing as everyone else – to make certain that the drug in question is appropriate for the condition it's treating.

But Pfizer used misleading selling tactics out of the Purdue playbook to market gabapentin to doctors for conditions that it was not or not yet approved for, and that there was little-to-no evidence that it was medically effective in treating. This prompted a lawsuit and in 2004 Pfizer pleaded guilty to improperly marketing Neurotonin; the company ultimately agreed to pay $325 million in a class action lawsuit. In the case of Lyrica, Pfizer was again accused of marketing the drug for off-label treatment of neuropathic pain before approval and, in 2012, paid out $43 million to settle claims in 33 states.

Ultimately, however, Pfizer's misdeeds didn't really touch their bottom line – Lyrica remained its flagship brand through 2018 and the highest-selling neurology brand for years. Much of those sales were from prescriptions for a wide variety of nerve pain, on the theory that if it worked for some, the mechanism could be the same for other conditions.* It also didn't do much to curb prescriptions of gabapentinoids – the number of people taking gabapentinoids

* What did worry the company was that time was running out on their patent, meaning Lyrica would lose out to cheaper generic versions; Pfizer's efforts to block the expiration of the patent failed. After US exclusivity ended in 2019, sales of Lyrica declined by half.

more than tripled between 2002 and 2015. And with the opioid crisis reaching its zenith soon after, that number just kept rising; prescriptions for gabapentin rose from 39 million in 2012 to 64 million in 2016.

But here's the real kicker: gabapentinoids don't really work all that well, and certainly not for the wide range of pain – chronic, nerve, acute – they're routinely prescribed for. According to a 2019 Cochrane review of existing evidence, pregabalin can offer a moderate reduction – as in, at most, 50 per cent, for a middling percentage of sufferers – in certain kinds of neuropathic pain, specifically pain related to shingles or diabetes. There is no evidence that it works at all in CNS pain disorders, and it has no benefit for individuals dealing with chronic low back pain or sciatica.

In 2019, two American doctors reviewed the existing evidence for gabapentinoids' effectiveness in treating a wide variety of pain conditions for a report in *JAMA*. The pair found that gabapentinoids are routinely prescribed not just for those pain conditions they were approved for, but also everything from migraine to osteoarthritis. However, they concluded, most review articles and guidelines 'overstate' the drugs' effectiveness – in many studies, patients taking the drug versus taking a placebo saw a less than one point difference on a 10-point pain scale. Dr Christopher W. Goodman of the University of South Carolina School of Medicine, one of the review's authors, told columnist Jean Brody of the *New York Times*, 'There is very little data to justify how these drugs are being used and why they should be in the top 10 in sales. Patients and physicians should understand that the drugs have limited evidence to support their use for many conditions, and there can be some harmful side effects, like somnolence, dizziness and difficulty walking.' Not only that, Goodman noted, gabapentinoids can also be addictive. Since then, a study published in December 2019 found that gabapentin, as well as another drug commonly used to treat chronic pain, the muscle relaxant baclofen, are linked to increased suicide attempts.

Gabapentinoids capture a lot of what's going wrong in trying to address chronic pain pharmacologically, and what has long gone wrong. As Brody wrote, 'The gabapentinoids are symbolic of three currently challenging problems in the practice of medicine: a deadly national epidemic of opioid addiction prompting doctors to seek alternative drugs for pain; the limited training in pain management received by most doctors; and the influence of aggressive and sometimes illegal promotion of prescription drugs, including through direct-to-consumer advertising.'

As Dr Art Van Zee, a Virginia doctor who watched opioids devastate his small Appalachian town, wrote in an article for the *American Journal of Public Health* in 2009, 'Controlled drugs, with their potential for abuse and diversion, can pose public health risks that are different from – and more problematic than – those of uncontrolled drugs when they are overpromoted and highly prescribed.'

Too much of what determines the division between 'controlled' and 'uncontrolled' drugs is driven by political agendas, by economic concerns over safety concerns, and in some cases by xenophobia, bigotry and bias; too little is driven by science. This paradigm disadvantages all of us, not least because so-called 'good' drugs, like opioids or gabapentinoids, are promoted to the point of addiction while the potential therapeutic and recreational benefits of 'dangerous drugs' are overlooked. We have never been great at drawing lines between drugs that are medicinal and those that are recreational because those lines are impossible to draw. That substances can be – and often are – both is a regulatory nightmare, but it's also had a serious impact on research. Case in point: marijuana.

Medical marijuana is hugely popular now, but for most of the 20th century, 'reefer' was the stuff bad kids used to get high and become a burden on society (see the 'drug-crazed abandon' of 1936's *Reefer Madness* for further details). Recreational use of cannabis was evidence of recklessness, irresponsibility and lawlessness; its use – like cocaine and, later, crack – was often attached to people of colour and immigrants to demonise them.* Due in no small part to the negative propaganda linking pot to 'dangerous' people, as well as the eventual criminalising of the drug, the identification and study of cannabinoids was slow to come. In fact, the psychoactive compound in cannabis,

* The popularity of the word 'marijuana', as opposed to 'cannabis', may be traced back to anti-cannabis campaigners' efforts in the 1930s to link the drug to the Latinx community in America, thereby further stigmatising it and them.

phytocannabinoid tetrahydrocannabinol, or THC, wasn't isolated until the 1960s.

The potential for cannabinoids in the treatment of pain-related ailments has only been taken seriously in the last 10 years or so. Studies show cannabis is helpful in managing anorexia, seizures, pain related to fibromyalgia, peripheral neuropathic pain, glaucoma and cancer; evidence suggests it works by binding to our endogenous cannabinoid receptors in our PNS and CNS. Cannabidiol, better known as CBD, is another compound found in the cannabis plant; it lacks psychoactive properties but is a safe and effective analgesic, and is a mild anti-inflammatory and anxiolytic, meaning it decreases anxiety. It is not addictive, and can be used to treat many of the negative side effects of opiates, including nausea and loss of appetite.

As with any drug, there are risks associated with the use of marijuana, most of them associated with chronic use and the method of delivery. Chronic use – a definition of which has yet to be standardised – shows a decrease in grey matter, especially for those who start while young. Long-term use is also related to a decline in executive functioning, information retrieval, learning, abstraction, motor skills and verbal abilities. Smoking it is obviously associated with lung disease, and it has been linked to cardiovascular disease, acute pancreatitis and cannabinoid hyperemesis syndrome, a cyclical vomiting disorder. There are also the risks that accompany using any mind-altering substance, such as workplace injuries, drugged driving and late-night takeaway.

But because cannabis is only legal in four countries across the globe, and it is still illegal at the federal level in the US, it's pretty difficult to research in the lab. And while the Drug Enforcement Agency is considering other growers, as of 2019 the University of Mississippi remained the only farm permitted to grow cannabis for research purposes, and they are capped at 1,000 pounds of cannabis each year. What this means is that we just don't have enough valid, reliable, replicated research on the costs and benefits of cannabis.

In fact, the biggest finding from a recent study, 'Cannabinoids for the treatment of mental disorders and symptoms of mental disorders: a systematic review and meta-analysis', published in *Lancet* in the autumn of 2019, is the confirmation that the data we have is insufficient to draw conclusions on the relationship to several mental health disorders. The authors constructed a careful analysis, controlling and accounting for as many differences between studies as possible, and noting the potential bias due to the fact that many of the study populations were based on disease diagnoses, specifically cancer and multiple sclerosis. With these caveats in mind, they failed to find solid empirical support for the use of cannabis in the treatment of depression, ADHD, Tourette's symptoms or psychosis. They did find that THC-CBD significantly reduced anxiety symptoms in seven studies, and one randomised, controlled trial looking at PTSD and cannabis that 'found a significant benefit of pharmaceutical THC-CBD compared with placebo in improving global functioning and nightmare frequency, and no significant effect on sleep quality'.

Drugs aren't all bad

The thing is, drugs – even opioids – *do* help people and relieve human suffering. We just need to get better at managing them.

Pharmacological interventions are often flawed, but they remain the most frequently deployed treatment, followed by surgery, for pain. But we need to know more about how they actually work. Increasingly, pharmacological researchers are looking to tailor their drugs to precisely what's needed and no more – the outcome of pain treatment should not involve a permanent weakening of or damage to our pain processing systems and certainly not to other essential organs. All this new knowledge and research should mean we can start sifting through some of the unintended consequences of our drugs. Still, a big part of the problem is that complex mechanisms and definitions are boiled down to slick marketing campaigns. These messages are not just oversimplified but also inaccurate and they leave patients and often doctors misinformed and with unrealistic expectations.

Right now, finding new pain treatments is like entering a labyrinth of indeterminate size. There are multiple directions facing us, and the further we explore, the more we'll find. There have already been a few dead ends and there will be more. It's exhilarating and, if we're honest, a little bit disappointing because, as we've seen over and over again, what seems like a miracle cure almost never is. But the history of drugs is also a history of experimentation, one that highlights the curiosity, creativity and innovation of the human mind and the limitations of the human body. Drugs are exciting; finding substances in nature or synthesised in the lab that can elicit feelings we've never had and mute the ones we don't want is magical. And few of us, if given the chance, would choose to live in a time before modern medicine. Yet, as we discovered or invented more drugs to heal ourselves, the possibility of a pain-free existence seemed like a realistic goal; we threw all the innovation, resources and research we had at developing new drugs, stronger drugs, faster-acting drugs. And in the wake of exciting discoveries and the rush to use science to cure everything, we neglected the consequences – and the alternatives. We neglected the fact that not all pain is bad, not all pain should be silenced, and that pain can be a source of strength. And we failed to recognise that there's a lot we can do to transform our experience of pain that doesn't rely on what's found in a pill, gas or bottle.

CHAPTER FOUR

Kids Need Pain – And So Do the Rest of Us

It was a warm, sunny day, only a few weeks into first grade and Thea had a plan. She'd never really played on the monkey bars – in kindergarten the year before, she'd been told they were off-limits for kids her age – but she wanted to show her five-year-old brother James, who was just starting at the same school, this 'cool trick' she'd thought of. Also, it had always seemed a bit unfair that she was supposedly too little to play on the bars. So this was her plan: she was going to climb up the metal ladder of the climbing frame, then jump to the first monkey bar and grab it with her hands. From there, she would pull herself up on top of the bars, run across the top, then swing down on the other side and climb down the ladder, like a ninja.

Thea climbed up the ladder. She steadied herself on the top bar. Then she jumped, hands outstretched for the bar. She missed.

'I meant to grab on one of the monkey bars but I didn't ... I couldn't because I was already falling,' she explained.

Thea plummeted to the ground, landing hard on her right arm and wrist. 'Then I couldn't speak for a while and then I could again and I said, "Owwwwwww," and I started crying,' she said. Her right arm was singing with pain and she couldn't move her fingers. The teacher on duty told another child to take Thea to the school nurse's office; the nurse saw that the arm would need to be x-rayed and immediately phoned Thea's mother, Julia. Thea waited in the nurse's office with an ice compress on her arm and her brother James beside her. When her mother arrived, she gave her 'lots of hugs,' Thea said. 'And then we were off to the hospital.'

Julia was at work when she got the call from Thea's school. The school nurse was calm, but direct – Thea absolutely needed to get to the hospital, how soon could Julia arrive?

'Oh god, shit,' Julia thought. She was worried about Thea first, of course, but the mental calculations weren't far behind: who could get there first, her or her husband, Duwayne? What work would have to be put off? How much was this going to screw up their heavily scheduled lives? They decided Julia would take an Uber to their home, then drive to pick up Thea in the family car; it took her around 45 minutes. Thea wasn't crying when Julia arrived, just sitting quietly with James and looking pale, her wrist in her lap. They all piled into the car; Julia asked Thea if some music might make her feel a little better and she requested Desmond Dekker. Over the vintage reggae beats of 'Israelites', Thea peppered her mother with questions. 'Do you think I'm going to have this memory forever? Do you think it's really broken? Am I really injured?'

In fact, Thea was 'really injured'. Her wrist was broken in two places; after a short wait at the hospital's emergency room, an x-ray revealed that Thea's bones would need to be repositioned and a cast put on. Staff administered a sedative and, Thea said, 'I fell asleep and they put my bones back in place.'

When we spoke to her two weeks after the accident, she was sporting a huge, dark blue cast that went past her elbow and made her small frame look even tinier. Like most six-year-olds, she can't stop making faces at the Skype camera while she tells us the story of her big fall and her broken arm, her curly brown hair in a wispy halo around her impish face. When she grows up, she's going to be a poet. Her favourite poet, she says, is herself.

The cast, she told us, widening her big brown eyes, is made out of 'fibreglass'. It is hard to sleep in, she said, but, as her mother pointed out, it certainly wasn't keeping her from running, playing, throwing her body around with the same kind of abandon. Asked whether she'll be attempting the monkey bars trick again, Thea said she might wait a little longer – at least until her cast comes off.

'When bones heal, they heal stronger,' Thea intoned. Poetically.

Thea isn't exactly right. Though bones, like muscles, weaken without use – this is why astronauts who have spent a long time in space, without gravity to work against, come back to Earth with decreased bone density – there is no evidence that breaking them causes them to heal stronger. But what about Thea's emotional bones, the internal structure that holds up her resilience, sense of self, her character and self-worth?

The function of pain in childhood is multi-layered, of course, but it has two big roles. The first is to teach us not to do dumb stuff that could kill us. We learn from every time we let the match burn down to our fingertips, or accidentally shave off a chunk of our thumb trying to 'whittle' a piece of wood, or jump off the monkey bars and miss. Those experiences and the resulting pain teaches us that risks don't always pay off, or, in some cases, give us information about how to do them better next time.

It's such a good, useful system, that researchers have tried to replicate it in non-living things. Johannes Kühn's robot is little more than an arm with articulated, motorised joints and a single pointer finger. But the cool thing about this disembodied arm is that, unlike other robot arms, it can 'feel' pain. Or something like it. Kühn is a robotics PhD candidate at the Leibniz University of Hanover in Germany; his artificial nervous system helps robots to 'feel' pain in order to keep themselves out of harm, opening up the potential for prosthetic limbs that act a little more like the ones they're replacing. The robot arm employs a biotech sensor on its fingertip – its only finger – that can register sensory inputs, specifically temperature changes and pressure, and can even differentiate degrees of 'pain'.*

Robots aren't people, of course, and the function of pain in actual humans isn't merely corrective. The second vital role of pain in childhood is to teach our fledgling brains that we can survive our own bad decisions.

The story of Thea's broken arm – the how and why and what happened after – is the kind of painful childhood experience that

* Kühn isn't the only researcher trying to mimic biological systems in robots. In August 2019 a South Korean research team announced they had developed an electronic skin that could detect heat and pinprick sensations. The technology has obvious applications for use in human prosthetics, but Professor Jae Eun Jang, head of the team that worked on the skin, suggested another use: controlling our future robot overlords. 'If robots can also feel pain, our research will expand further into technology to control robots' aggressive tendency, which is one of the risk factors of AI development,' he said.

can lead to healthy physical, emotional and social development. As far as broken bones go, her fracture was fairly straightforward and won't have any lasting effect on her physical ability – just a quick reset of the bone, keep it in a cast for eight weeks, pop off the cast, and back to normal. And everything around the break itself – from Thea deciding to try that 'cool trick', to how the school staff and her parents treated her, to how she considers her own ability after the incident, and how she shared the experience with us as she made faces on Skype from her kitchen table – will have a lifelong and likely positive impact.

But when children don't have opportunities to make decisions that might result in pain, they never get the chance to learn from that experience and to make better predictions next time. Emotionally, they never learn that they can manage and withstand pain when it occurs, reinforcing a cultural message that children are fragile and the world is dangerous. Physiologically and neurobiologically, their expectations of painful experiences will not match reality and this mismatch can lead to increased pain later on. Research demonstrates that the pain we feel in childhood and how it's treated goes a long way towards shaping our processing of painful events later in life.

At the 2016 mid-year meeting of the American Psychosomatic Society, Dr Tor Wager of the University of Colorado explained how our expectations govern our experience of pain. Wager, whose pain signature theory we discussed in Chapter 1, makes a compelling case that expectations establish not only our perception of pain in the moment but also in future experiences with pain. Basically, our ventromedial prefrontal cortex (VmPFC), the neural area heavily involved in decision-making, calculates expectations based on what we know now and our prior experience. Once expectations have been set, the VmPFC sends information to the periaqueductal gray, which, when activated, deploys our endogenous painkillers. If expectations do not match experience, our body does not provide the pain relief we need in the moment and we feel worse. And when we expect to feel good all the time, and lack experience dealing with pain because we (or our parents) have so effectively shielded ourselves, we can't make good predictions.

Kids need pain. But it has to be the right kind of pain, in the right context and with the right attention to their needs, both physical and emotional. Some of this is easier said than done: kids are pretty good at getting hurt when left to explore and adventure. Everything else can be a little more of a challenge.

How 'better safe than sorry' is actually making us sorry

Thea was not always as ambitious when it came to physical activity, and neither were her parents anxious to push her. Thea was born six weeks premature; her birth remains the scariest parental memory for Julia and Duwayne. 'After she was born, she was so little,' said Julia. 'I worried a lot about her being fragile or being sick.' Though Thea was not medically frail, her motor skills were slightly delayed. 'Other babies could do what she couldn't. Like walking. She couldn't run for a long time, couldn't figure out the mechanics of that, but other kids could,' Duwayne explained.

As she grew, Thea's own relationship with her body was more guarded and she was more cautious than other children – her brother was hurtling himself down slides at two-and-a-half that Thea was afraid of at four. Wanting Thea to grow up feeling confident and powerful in her own body, Duwayne and Julia sought guidance from Thea's preschool teacher, Jarrod Green. That guidance turned the four-year-old who was afraid to join her younger brother on a slide into the six-year-old plotting ninja moves on the playground monkey bars. 'I blame him for her fearlessness of jumping from any height,' Julia said, laughing.

Green is the author of *I'm OK! Building Resilience Through Physical Play* and director at the Children's Community School (CCS), a preschool in west Philadelphia whose explicit mission is to empower 'children to engage their whole selves in education'. In practical terms, this means a lot of messy play, a lot of wardrobe changes and a few bumps and bruises. The school inhabits one of the area's many old churches, redeveloped in an era of declining parishioners and rising property values. Inside, it's classic nursery – clean, bright colours, toys, children's art, foam mats for tumbling on, tiny chairs. Out front, however, is what looks like a derelict building site: construction cones, car tyres, milk crates, metal shovels, a large silver trough and utility tents scattered across well-trodden mud. That's the playground. (And believe it or not, it's fully insured.)

Green, who earned his masters in early childhood education from San Francisco State University, is among a growing number of educators who recognise the critical importance of physical play in children's development. Implicit in the promise of physical play is the kind of risky play, preferably outdoors, that carries a chance of getting hurt. This is crucial to what the school describes as its 'play-based, emergent curriculum'. It's not easy to define exactly what a 'play-based, emergent curriculum' is, but basically, the approach is built on the principle that powerful learning – not just about motor control but also about self-awareness, resilience, empathy and even justice – emerges, if you will, in those moments where we are testing the boundaries of our abilities.

Teaching the practice of resilience is central to Green's philosophy as an educator. 'Resilience' became a kind of buzzword in the last decade on the backs of several studies and think-pieces demonstrating that a lack of it was harming the mental health of children in Western society. There are many definitions of resilience, a fact that has significant consequences when it comes to comparing study outcomes. But in this case, Green's definition of resilience is similar to the one adopted by most researchers today: 'a dynamic process encompassing positive adaptation within the context of significant adversity'. Critical in this definition is the departure from defining resilience as a personality trait, as something one can 'have' and deploy equally across all contexts. Instead, resilience as a process captures the active nature of adapting to challenges in ways that result in positive growth. And as a process, we can learn it, practise it and appreciate that we'll be better at it at some times than others.

Green offers an example. During his first year at CCS, he encountered a child who was afraid of climbing. 'There was a moment where he was trying to climb over a slide and got stuck with one leg down the ladder and one down the side and he became very, very fearful,' recalled Green. '[I was] being very conscious in the moment of like, "Oh, this is a resilience moment." He is very scared and he could either learn, depending on my response, he could learn, "Right now, my fear is valid and I need to be rescued" or he could learn "I have competencies, I'm safe."' Green coached the child through the moment, reminding him in a calm, natural tone of voice that he had been successful in situations like this before, and pointing out resources he could use. But critically, he didn't rescue the child. The child was fine, of course, and became more comfortable with climbing and with his own abilities.

'The thing you fear happening is very unlikely to happen, and should not be the thing driving your decision-making,' explained Green. Instead of focusing on caution and safety, caregivers ought to focus on fostering opportunities to learn resilience. At CCS, they're empowering children by trusting them to explore, interact and problem-solve in a rich environment that is *as safe as it needs to be.* This sounds like common sense, but it's actually a critical deviation from the trajectory of the last half-century, which has been to make all child-focused environments as safe as possible. And we're now learning that there's such a thing as too safe.

The push to build 'safer' playgrounds – impact-absorbing surfaces, height restrictions, guard-rails – has reduced arm fractures like Thea's as well as more serious injuries, including traumatic brain injuries. That's significant. But it has also made play less fun. And when kids don't play, their emotional and physical development suffers.

We recognise that it seems like we're pitting the value of play against the value of physical safety. But what started as efforts to make playgrounds safer evolved by the 21st century into eliminating even the whisper of risk. Schools and parks increasingly opted for the bright, multicoloured, plastic composite play structures: the now-familiar climbing station, tower and slide combination, purchasable in an off-the-shelf package. These structures – even with their fluorescent colours – are boring. If the words of kids themselves aren't enough, there's now research to back it up. In 2015, Studio Ludo, a non-profit dedicated to 'building better play through research, design, and advocacy', studied 45 playgrounds around London. They then compared their own data to US data gathered under the RAND Corporation's National Study of Neighborhood Parks. Comparing London playgrounds to those of similar size and in similar areas in

San Francisco, New York City and Los Angeles, they found that UK playgrounds see 53 per cent more visitors, including 14 per cent more adults, and children and teens were 16 to 18 per cent more active.

Why? The report suggested that London playgrounds were successful because they blurred the boundaries between the play areas – themselves filled with sand, swings, places to climb – and the grassy park areas, creating adult-friendly spaces for 'dwelling' and relaxing. They had fewer off-the-shelf components and were more organic, taking advantage of what Studio Ludo called 'play affordances', natural elements in the built environment such as large boulders, hills, trees and logs. Perhaps most striking was that the UK playgrounds, despite inviting what Studio Ludo called 'riskier play', enjoyed lower rates of injury than US playgrounds. The study also noted that the playgrounds in the US are more expensive – the average cost per square foot in the UK is $30, compared to $48 in the US, a difference largely attributed to surfacing materials (grass, sand and tree bark in the UK; poured rubber and rubber tile in the US).

The difference between the playgrounds, however, isn't just in what kids are running and sometimes falling on. It underscores a more significant cultural division. 'The U.S. seems to have reached "peak safety". We have created a nation of overly expensive, homogeneously safe, and insidiously boring play spaces,' concluded Studio Ludo. 'Our injury rates demonstrate that these spaces have unintended consequences. In pursuit of fun, children are using play structures in unintended ways, falling on surfaces too expensive to maintain, and are not moving enough, becoming too weak to play without injuring themselves.'

Reflecting an effort to balance safety and fun, government agencies in Britain, Australia, Sweden and Canada have all adjusted their playground safety guidelines; they are now based on a calculation of risk and benefit, rather than risk alone.* That is not the case in America, however, where discussion remains centred on the elimination of

---

* A systematic review into playground accidents published in 2019 in *Canadian Family Physician* appeared under the title 'Don't take down the monkey bars' and concluded, 'Although playgrounds are a common location where pediatric injuries occur, these injuries are relatively low in frequency and severity.'

all risk with the driving goal of mitigating liability rather than promoting engagement. That's not only impossible, it's also the wrong target. Risky play is just what kids need and what they want.

'In modern western society, there is a growing focus on the safety of children in all areas, including situations involving playing. An exaggerated safety focus of children's play is problematic because while on the one hand children should avoid injuries, on the other hand they might need challenges and varied stimulation to develop normally, both physically and mentally,' wrote Ellen Beate Hansen Sandseter, a researcher from Queen Maud's College in Norway, in a 2011 article for *Evolutionary Psychology*. 'Children test possibilities and boundaries for action within their environment through play, most often without being aware that this is what they are doing … The rehearsal of handling real-life risky situations through risky play is thus an important issue. Paradoxically, we posit that our fear of children being harmed by mostly harmless injuries may result in more fearful children and increased levels of psychopathology.'

Risky play describes thrilling and exciting forms of play that can include the possibility of real physical injury; risky play is not, however, handing your kid a pack of matches and saying, 'Have fun.' It's about creating spaces for children to climb and run, to work with (potentially) dangerous tools, to navigate around natural elements like water and fire, to even get lost; if you were ever a Scout or a Girl Guide, this kind of play is baked into the ethos of these organisations. The idea is that it is empowering for kids to explore and test their agency, to recognise hazards and risks, to learn – through experience – cause and effect, and it is essential for healthy physical, social and emotional development.

Science backs this up. Two systematic reviews of research on the relationship between risky outdoor play and children's health concluded that the benefits heavily outweigh the threats. Kids play longer and are more physically active – running, jumping, getting their heart rates up – when they are engaged in risky, particularly outdoor, play. Studies show improvements in cholesterol levels, blood pressure, body composition, bone density, cardiorespiratory and musculoskeletal fitness. Other studies also show that risky play fosters motor skills, prosocial behaviour, independence and conflict resolution. An experimental 2015 study by Ann Lavrysen, from the Katholieke Universiteit Leuven in Belgium, found that children involved in a 14-week risky play intervention showed improvement in the following areas compared to both their own pre-intervention scores and the non-intervention group: self-esteem, risk detection, competence and a decrease in conflict sensitivity. These findings also suggest that when risky play isn't available, children are more likely to develop problematic behaviours around hazards, phobias and overwhelming fears.

But what about when a child *does* get hurt, say, falling off the monkey bars? Then what? According to researchers from The Hospital for Sick Children, a major paediatric teaching hospital in Toronto, Canada, parents and caregivers can support children in pain through the 'three Ps', a triad of psychological, physical and pharmacological approaches. The physical and pharmacological approaches involve injury- or illness-appropriate interventions, such as a cold compress or paracetamol.* The psychological approach can be a bit more nuanced – they recommend following the ABCD method of managing pain: 'A' is for assessing your own anxiety – a calmer parent makes for a calmer child. 'B' is for 'belly breaths', which slow breathing and heart rate, dampening down sympathetic nervous system over-activation; if you do it while your child is in your arms, it will also calm them down. 'C' is for cuddle or close physical contact, which studies have demonstrated reduces stress. And 'D' is for distraction, especially after the peak moment of pain has passed.

All of these are important, but distraction is one of the most powerful tools in caregivers' kit. For example, a 1999 study found that children coped with the pain of vaccination jabs better when they were distracted – even better than when receiving topical anaesthetic. Dwelling on a painful experience underlines that experience and gives it outsized meaning to the child, so distraction can help turn the focus off. But – and this is why we've taken the

* Never underestimate the power of a well-timed plaster in mitigating children's acute nociceptive pain. Especially if that plaster bears the image of a beloved cartoon character. In the event cartoon-character plasters are in short supply, a simple smiley face in marker will do. This is *science*, people.

time to dig in to this – distraction doesn't mean ignoring or diminishing pain. Not only will this kind of attitude exacerbate their perception of pain in the moment, but recent research also demonstrates that dismissing children's pain has lasting effects on their well-being into the future. A study published in 2018 in *The Clinical Journal of Pain* found that one in 10 of the nearly 2,000 young adults surveyed reported an incidence of pain dismissal when growing up; half of these incidences involved a parent or medical provider. These were the most distressing findings: 'Often, this experience was associated with a lasting sense of hostility toward the person who dismissed their pain, as well as anger, self-directed negativity, damage to the relationship, and feelings of isolation', researchers found. Notably, the incidence of pain dismissal rises precipitously among children with chronic pain conditions – more than 40 per cent of children with chronic pain conditions described others denying or diminishing their report of pain.

Thea's broken arm is a great example of a painful experience with a positive outcome. Thea took initiative, a risk, and jumped, exploring not only her physical ability but also taking a leadership role – she wanted to show her brother James a new trick *and* challenge the logic of why the older kids could play on the monkey bars, but she couldn't. When she missed the monkey bars and broke her arm, she was not made to feel like a failure or that she was 'wrong' to have attempted it. In the immediate aftermath of the accident, the school's nurse and Julia remained calm; Julia gave Thea 'lots of hugs' and helped distract her by playing music she liked. Her pain was neither catastrophised nor diminished, and her questions and concerns – 'Am I really injured?' – were met with consideration and age-appropriate responses.

That Thea was ready to try the monkey bars again – well, when her cast came off – implies that this experience and the support around it bolstered her skills of resilience. As her story demonstrates, you can't cultivate skill or resilience without testing limits. Children can learn the relationship between action and outcome from reading or hearing a story or through observation. But the skills of resilience and of knowing one's body are best learned through experience, even if that experience is falling off the monkey bars. They can be told dozens – and dozens and dozens – of times* to not do that dangerous thing because it's dangerous, but they often do that dangerous thing anyway. And they not only learn that you were right, they also learn what that kind of danger feels like and add a whole bunch of new sensory data points to their growing inventory of experiences.

The next step is teaching them how to identify and talk about those sensory experiences.

How to talk about pain

Big belly breaths and cartoon-stamped plasters are efficient tools in the aftermath of injury, but the most powerful of them all is language. In Chapter 1 we introduced the theory of constructed emotion and how we learn to make sense of our physiological sensations in a social context in part by learning language in interaction with the people around us. Words are mental concepts linked to specific kinds of sensations; they become 'shortcuts' that help us get what we need in a less taxing, less tear-filled manner. The more shortcuts we have, the quicker we can get to places and the more alternative routes we have should one be closed off.

When it comes to working through the cacophony of sensory signals following a fall, words help kids get back up and keep going. But 'pain' and 'hurt' shouldn't be the only words we use to describe times when we feel less than good. This is one of the first lessons acrobatics instructor Zach Fischer teaches to his students. Fischer is Jarrod Green's brother, and we knew we wanted to talk to him after Green said his approach to teaching kids to navigate preschool is a lot like his brother's approach to teaching older kids to stand on their heads. Fischer describes their childhood as typical – climbing trees, pretending to be Spider-Man – but he assumed by his mid-twenties that he was too old to learn any real superhero skills. When he moved back to California at age 26, however, he discovered Athletic Playground in Emeryville, an alternative gym that encourages people of all ages to move in new and exciting ways. Ten years later, he

---

* So. Many. Times. *So many.*

teaches kids and adults acrobatics and parkour, which is basically getting from point A to point B as quickly and creatively as possible or, as Fisher says it was called when he was a teenager, 'Hey, get down from there!' Fisher's experience of teaching kids in activities where pain is guaranteed has given him insight into how kids make sense of what is happening in their bodies. Especially when that handspring turns into a belly-flop.

'They experience something unexpected that registers as pain, but I would say 80 per cent, 90 per cent of the time, they are not hurt at all. There's no lasting effect other than the internal emotional sensory processing that has to go on, to be like, "Oh, I just got a bunch of red flags in class,"' he explained. Those red flags can be startling, and it's Fisher's job to help kids navigate the sensations overwhelming their young systems. Crucially, Fischer doesn't start by asking how much it hurts, or even where, but rather by acknowledging the shock of the experience. 'We're like "Wow, that was surprising wasn't it?" and they're like "Yeah!"'

He then helps them check in with their body, asking: '"OK, well, what do you feel in your body?" And they're like, "Well, I feel this … stinging sensation here." And then "OK, what else do you feel?" and they're like, "An impact here."' Fischer is teaching his students to bring awareness to the sensations, and to use different adjectives to describe what they're feeling. The more words – the more 'flavours of pain' as Fischer puts it – the better. 'When it's just pain, there's no delineation, no way to help yourself use that information,' he says. 'It's just one big scary signal and I've got to avoid everything that's going on now.'

Only after they describe the sensations does ask students to rate its intensity on a scale of 1 to 10. Once they've gone through that process, he asks them to count to 10, take some deep breaths and check in with their feelings again. This process of acknowledging a sensation, followed by localisation and description and then a reflection on intensity, is helping kids not only gain a greater understanding of their body but also connect distinct sensations to mental concepts, creating the shortcuts that will allow them to work through similar sensations in the future. In the short term, Fischer is distracting kids from pain by helping them focus on sensation. In the long term, he is teaching these kids emotional granularity, and it's a powerful skill to learn.

Emotional granularity refers to the ability to differentiate between unique emotional instances and label them with specificity. Emotions are not universal states that are experienced the same way from person to person, nor are they discrete – anger isn't just anger, it can be anger plus frustration, anger plus sadness – but building and expanding our own emotional dictionaries carries big benefits. Studies show that those with a high level of precision and awareness of their emotions demonstrate better emotional regulation skills, including adaptive coping and adjustment, greater resilience in the face of stress, less aggressive behaviour in upsetting situations and the ability to prevent emotions from biasing moral judgements. Higher emotional granularity is also associated with overall better well-being, lower rates of alcohol and drug abuse, going to the doctor and using medication less frequently, and spending fewer days in the hospital. On the other hand, low granularity, especially when it comes to differentiating between negative states, is associated with poorer emotional regulation, social anxiety and major depressive disorders.

Research measuring differences in brainwave activity with EEG offers insight into how and why high granularity is so useful. In a study published in *Frontiers in Neuroscience* in 2017, Ja Y. Lee from the University of Wisconsin-Madison, along with co-authors Kristen Lindquist and Chang Nam, examined differences in brainwave activity between individuals as they reacted to emotionally affective images. Neural activity suggested that individuals with high granularity had better 'access to executive control resources and a more habitual processing of affective stimuli'; they could reach conceptual knowledge more quickly and with less cognitive effort, enabling them to process what was happening and take steps to regulate their emotions more quickly. 'Low granularity individuals may thus have early reactivity to affective stimuli, but do not then engage conceptual resources to help make meaning of their reactions to affective stimuli,' they concluded. Without a conceptual understanding of the sensations happening, regulating the affect resulting from them is difficult. In other words, it's like an alarm is going off somewhere but having no idea where it's coming from, what it means or how to make it stop.

The ability to differentiate the flavours of painful sensations is important, but so is what's called emotional diversity or having lots of different types of emotional experiences. Emotional diversity isn't concerned with the intensity of emotional experiences but rather the range and frequency of different feelings over a given period of time; having a diversity of positive experiences is especially beneficial. The wisdom of 'live life to the fullest' is more than just a good motto: in a 2018 study asking participants to report on their emotions for 30 days, researchers found greater positive emotional diversity was related to lower inflammation, independent of age, gender, anti-inflammatory medications, BMI, medical conditions and personality. Another study, this one of more than 37,000 respondents, found that emotional diversity is an independent predictor of mental and physical health: high emotional diversity is associated with fewer doctor visits and lower rates of depression.

Just as practising a handspring over and over again is necessary for mastery, over time and with work, the pathways of managing physical sensations by employing unique mental concepts and descriptive words become automatic. Fisher has seen this work with students. A boy who had been training at the gym broke his arm during an exercise. 'He was clearly experiencing pain, but he was like, "I think I'm hurt, I have this really sharp pain in my arm that's not going away,"' Fisher explained. 'He's not freaking out, he's undergoing that same process we taught him over the years. But now there's a real problem and so he's able to look at it and be like "Actually, I think I broke it."'

Encouraging kids to develop a rich vocabulary for all the different sensory experiences they will have not only makes for an exciting childhood but also gives them the tools they need to navigate through pain and stress and to celebrate the joy and wonder in the world. To do that, they need to have all kinds of experiences, including painful ones.

Now the tricky part – getting parents to let them.

Paranoid parenting hurts

Most of the parents who sign their kids up with a class taught by Zach Fischer or who walk through the doors of his brother's preschool are doing so because they believe in the benefits of risky play for their kids. But as both brothers point out, knowing, believing and wanting to do something does not always mean you actually *can* do it.

According to Fischer's brother, Jarrod Green, the tough conversations he has with parents aren't when the child gets hurt, it's when parents grapple with the possibility that they might. Their commitment to his approach can only really be tested in the moment – for example, standing by as they watch their preschool child build and then attempt to climb a wobbling mountain of tyres. In those moments, Green and his staff are teaching the parent as much as the child. 'I'll be watching a nervous parent, or parent that is likely to step in, and I'll stand next to them and narrate something dangerous. And once I'm in conversation with them they can't just walk away, especially when they know I'm watching,' Green said, laughing. 'So I'll be like, "Oh look at Suzy over there. She's really up high, isn't she? I've been watching her, she's really been building up courage at this. I want to see what she does. I know she's a pretty competent climber. I think I'm gonna stay a little far back and give her room, but I'm watching."'

Green is both reinforcing the boundary with the parent and demonstrating his own thinking process – it's the kind of thing he does with the children at the preschool, but he says it works just as well with adults. To teach resilience, the parent has to be willing to cultivate and practise resilience themselves, through letting go and trusting. 'Part of it is trusting [children] to make better choices than we fear they will, and part of it is trusting that when they don't make good choices, it's not the end of the world, it's learning,' he said. The experience can also help parents confront some potentially difficult realities about their own relationship to risk, injury and pain.

But the hesitancy Green observes in his students' parents reflects a cultural struggle with letting our children take risks, with living in a society inundated with messages of fear and avoidance. When, for example, was the last time you saw a child under the age of nine walking somewhere, anywhere, alone? In the UK, 86 per cent of primary school children walked to school by themselves in 1971; according to the National Travel Survey, only 3 per cent of primary school children travelled to school alone in 2013. The survey also asked parents why they accompanied their children to school; the main reasons were traffic danger and 'fear of assault'. Meanwhile, the going advice in the UK is to 'never let a child under the age of 10 cross the road alone'.

In fact, in 2008, when Lenore Skenazy let her nine-year-old son find his way home by himself in New York City, she was branded a child abuser, a terrible mother, America's 'worst mom' (of course, writing about it for her *New York Sun* column might have had something to do with the volume of the backlash). Skenazy, then also running a blog called 'Free Range Kids', wrote that her son had been begging her for weeks to leave him somewhere and let him find his own way home. So she left him in the handbag department of Bloomingdales with a subway map, $20, a MetroCard and a couple of quarters. A few hours later, he walked through the door of their flat in Queens, 'ecstatic with independence'.

Skenazy was already evangelical about the right of children to grow with independence, but the experiment catapulted her into the mainstream. She briefly had a reality-TV show, showcasing parents who needed to be talked off the ledges of their own anxiety, and started a non-profit, Let Grow, dedicated to teaching parents and others how to trust children to make their own way in the world. Speaking to her now, she is incredibly energetic, talks fast and, well, sounds *right*.

'For years, I asked audiences, when I would give a lecture, for something that they did as a kid that they don't let their own kids do. And it was a pretty fun conversation,' she told us. She heard stories of climbing trees, riding bikes until the streetlights came on, that kind of thing; she then started asking people, 'What went wrong when you were out on your own?'

Mostly, she said, people got lost or got hurt. One woman remembered hurtling down a hill on her bike, hitting a patch of slick pine needles, and her handlebars falling off. She threw herself off the bike and landed in a bush, but she fixed the bike and pedalled home. Another guy remembered getting lost and hitch-hiking home. One guy, she said, was playing mumbly-peg (a game in which one player throws a pocket knife into the ground; the other player has to pull it out with their teeth) and a kid took a knife in the foot. 'They washed off his foot in the shower and covered it over in Band-Aids,' she said.

These experiences didn't destroy or scar them; if anything, they confirmed they could handle pain, danger, uncertainty. 'All of these experiences were either terrifying or painful or both. And they're very easily recalled and they're always a point of pride. What I try to impress upon the people I'm speaking to is, what if this low point was really a high point in your childhood? When dealing with pain, dealing with fear, dealing with unintended consequences of somewhat grave magnitude, is a way that you define yourself, the idea of not letting kids do anything because something could go wrong is the opposite of what you're grateful for.'

But for a parent, it's difficult to watch your children in situations where they might get hurt, in part owing to the cultural feedback that reinforces the narrative that children should never be in such a situation in the first place. However, just as we're watching our kids, they're watching us. Children model behaviour and learn how to cope from parents – a hesitant child often means a fearful parent.

As the recollections of Skenazy's audiences – and our own – reflect, we haven't always been so afraid of letting kids out of the house. In the 1980s, however, high-profile stories of abducted and murdered children fuelled 'stranger danger' campaigns; the response most kids heard when they asked to do something their parents wouldn't allow was, 'It's not that I don't trust *you*, it's that I don't trust *other people*.' And by the 1990s, paranoia had replaced caution.

'We've taken the possibility and moved it into the category of probability,' Skenazy said. 'We keep thinking that if something is not 100 per cent safe, like, if it's 98 per cent safe, then it's not OK.' At the same time, real probabilities of harm have actually decreased. Since 1990, according to the CDC, child mortality rates in the US have fallen by half (though the rate for Black infants is still twice that of white), a decline that has less to do with parenting and more to do with advances in medicine and technology. According to the UK Department of Transport, the number of child pedestrians killed or seriously injured by cars declined 84 per cent between 1979 and 2013 (from 12,458 to 1,980 and yes, quite possibly *because* there are fewer children on the streets). Our fears of child abductions are also largely unfounded: each year since 2010, fewer than 350 people under the age of 21 have been abducted by strangers in the US, according to the FBI. In 2016, child murders in the UK hit the lowest number since record-keeping began – a total of 36 children under the age of 16 were killed by another person that year Over-protective parenting can account for some of these declines – pedestrian accidents, for example – but not all.

But even as our environment became safer, our knowledge of the consequences of injury increased. Consider, for example, concussions. The number of playground falls stayed about the same from 1992 to 1997, yet hospital visits for children after playground falls increased in that same period. Researchers suggested that one reason might have been the increasing awareness around head injuries and concussion – injuries in the past that might have been shaken off are treated as more serious. Concussions are no joke, but though there has been an upward trend in US hospital visits over the past 10 years, the percentage of those under the age of 18 who are examined, treated and released has held steady at 95 per cent. As Jarrod Green, who has treated more than a few paediatric bumps, bruises, scrapes and cuts, said, 'Most injuries just aren't that bad.'

In general, we are not great at risk assessment, but living in an increasingly connected world has made the job of determining the real probability of harm even harder. The internet and social media mean we know more about the horrible, dangerous things happening in our community and around the world all the time. The threat response motivates us to act, and our increased knowledge of the world comes with an increased but utterly illusory expectation of control. We accept that bad things happen but have refused to accept that we are not always able to control when or where they do. And this produces anxiety, which is then directed at the things we believe we can control. Like walking to school alone.

Professor Jean Twenge, a psychologist at San Diego University and author of 2018's *iGen: Why Today's Super-Connected Kids Are Growing Up Less Rebellious, More Tolerant, Less Happy – and Completely Unprepared for Adulthood – and What That Means for the Rest of Us*, suggests that the way we raise children now – physically safer than they have ever been, by a number of metrics – is part of what is making them less happy compared to previous generations. Twenge claims that iGeners (defined as children born in the mid-1990s and later and who, crucially, have had access to social media and smartphones for much, if not all, of their lives) have been raised on a constant drip-feed of messages about physical safety. They are less likely to drink and drive, have sex at a young age, become parents as teenagers or take drugs. That is all good. But they are also less likely to experience anything outside that neat little rectangle and they rarely go out without their parents – 12th graders in 2015 were going out less often than eighth graders did in 2009. Not going out is not inherently worrying, but in stride with these changes is data showing teens are also dramatically less satisfied with themselves and their lives than previous generations were. They are the loneliest they've ever been since the Monitoring the Future survey started asking high schoolers how they feel in 1991. And teens' depressive symptoms – agreeing with statements like 'My life is not useful' and 'I do not enjoy life' – have skyrocketed since 2011, reaching all-time highs.

This kind of observed change over time is worth paying attention to, and Twenge's research is based on data stretching back to the 1970s, but there are a few caveats. First, any historical view of emotional states should be regarded with caution – it's not as if psychologists were handing out surveys on emotional well-being in 1890. Second, correlation isn't causation. There are a number of other changes – as she acknowledges – that are contributing to iGeners' diminished sense of well-being: this is the first generation not expected to live longer than their parents; millennials, the much-maligned generation just before iGeners, are the first to earn less than their parents; meanwhile, there's good old political division, threat of climate change, and, as we were writing this book, a global pandemic. Thanks to the internet, we know about all of this, so they'd be forgiven for thinking the future's not looking so bright and the only shades we are going to need are for social distancing.

At the end of her book, Twenge wrote, 'I've realized this: iGeners are scared, maybe even terrified.' Maybe they should be, but her conclusions also demonstrate that when we bubble-wrap our kids and protect them from risk and pain, we don't do them any favours; all we do is reinforce the message to them that they *need* to be bubble-wrapped.

Given all the actual existential threats facing this generation, that's a dangerous message. In order to better withstand the real threats coming their way, kids need more opportunities to experience pain and discomfort on a smaller scale, and as a result of their own decisions. Learning that you can handle pain is 'suddenly feeling like you have this superpower you didn't know you had,' said Skenazy. 'The only way you can discover it is when there is nobody to take the pain away.' Parents can't be there to protect them from all the slings and arrows that life will fire at them – being cut from the team, the mean girls, the falls from the monkey bars – so kids have to learn how to protect themselves. 'Those are all painful, but they do create this bedrock of recognising what you can deal with and that's a deep gift.'

Good pain, bad pain, and very, very bad pain

Pain in pursuit of a goal, pain as a natural consequence of play or life, pain that is acknowledged and met by caregivers, that's the pain that can lead to those superpower moments Skenazy described. Practising how to manage pain and adversity on relatively little things – a skinned knee, a broken wrist, a pocket knife through the foot – is practice for big things. These experiences build skills of resilience and coping that will help children confidently manage the stresses, both physical and emotional, that are inevitable in life.

These lifelong emotional implications rely on neuro-physiological processes that are themselves shaped by our experiences. As we talked about in Chapter 1, the systems that perceive and report sensory input, including nociception, develop in the weeks before full gestation, in the natural development of our PNS and CNS. But how these systems learn what to do with that information and how to send out signals that respond automatically and appropriately, takes experience. Another way to describe this process is as researchers from the Center for the Developing Child at Harvard University do in a working paper entitled 'The Science of Neglect':

> The architecture of the brain is composed of highly integrated sets of neural circuits (i.e., connections among brain cells) that are 'wired' under the continuous and mutual influences of both genetics and the

> environment of experiences, relationships, and physical conditions in which children live. Experiences 'authorize' genetic instructions to be carried out and shape the formation of the circuits as they are being constructed. This developmental progression depends on appropriate sensory input and stable, responsive relationships to build healthy brain architecture.

Basically, our genes and environment offer up the map of our mind and body, but it is our experience in it – the pain we feel and how it is treated – that forms the navigation system and sets the course. Which is why our early experiences with pain are so important.

Much of what we've talked about in this chapter so far has rested on a few big assumptions. We have largely been talking about parenting and caretaking in developed nations, parents who have a choice about what playground to frequent and whether to helicopter or not, parents who are not themselves harming their children. Children enduring pain as part of treatment for life-threatening diseases will have a very different perception of pain than children living with abuse, whose perception of pain will be different to children dealing with the aftermath of a traumatic accidents. Children in war-torn, deprived, rural or poverty-stricken parts of the world are not being denied opportunities for risky play; risky play may be their only option, if play is an option at all. Some kids are raised with the message, communicated through media, social environments and play spaces, that you are not safe unless we make you safe; others learn, quickly and accurately, that they're not safe at all. This is what we might talk about as bad pain, pain that is inflicted upon the individual and that feels outside of their control, unexpected, unfair. And its consequences are devastating.

Unfortunately, the scope and extent of the consequences related to early childhood adversity or trauma – because remember, pain is pain, emotional or physical – has largely been neglected by medical doctors. The oversight is not intentional: under the dominant biomedical model, primary care doctors treat the problem not the patient. But in 1995, Dr Vincent Felitti, founder and head of Kaiser Permanente's Department of Preventive Medicine in San Diego, started a not-so-silent revolution to bring attention to the impact of childhood adversity.

After years of observing the relationship between negative health outcomes in adulthood and adverse childhood events, Felitti and his team launched the Adverse Childhood Experiences study through the Kaiser Permanente's San Diego Health Appraisal Clinic. This clinic saw more than 45,000 adults each year and at the time was one of the largest free-standing medical evaluation centres in the country. Their goal was to collect data on disease burden and risk factors, quality of life, healthcare utilisation and overall mortality. Since this was the first study of its kind, Felitti and Dr Robert Anda designed a new questionnaire that pulled from several different standardised scales. Seventeen questions were organised into abuse categories (psychological, physical, sexual) and household dysfunction categories (substance abuse, mental illness, violence against mother and criminal behaviour). The response rate was impressive, yielding a final sample of 17,337 patients.

In 1998, Felitti and his team published their results in the *American Journal of Preventative Medicine* under the title 'Relationship of Childhood Abuse and Household Dysfunction to Many of the Leading Causes of Death in Adults'. Almost two-thirds of respondents reported experiencing at least one 'adverse childhood experience', or ACE, with a little over one in five experiencing three or more ACEs. There was a strong correlation among abuse categories, meaning if someone experienced one form of abuse or dysfunction, the chances they experienced multiple others was high. This in itself is critical because too often studies silo abuse and dysfunction, potentially missing the compounding effect of multiple adverse experiences.

But what was even more surprising were the results of how early adversity related to health risk factors, disease incidence, healthcare utilisation and mortality. As ACE scores increased, so did the prevalence and risk of smoking, severe obesity, physical inactivity, depressed mood, suicide attempts, alcoholism, use of illicit drugs, injection of illicit drugs and sexually transmitted disease. But so did the likelihood of being diagnosed with ischemic heart disease, cancer, chronic bronchitis or emphysema, hepatitis or jaundice, skeletal fractures and overall poor self-rated health. This was even after taking into account age, gender, race and education, factors known to have a big impact on disease burden.

What the ACE study found was a definitive correlation between adverse childhood experiences and a whole universe of negative health outcomes. Previously, no one had looked for these connections – which meant that no one had really ever taken that trauma into treatment consideration.

At this point, you might be wondering how much trauma we are talking about – what's 'bad' enough to cause these kinds of dramatic changes and outcomes? That's almost an impossible question to answer, in large part because pain or trauma is specific to the individual and the context. But consider the questions asked on the ACE screening questionnaire:

> 'Did a parent or other adult in the household often ... Swear at you, insult you, put you down, or humiliate you? or Act in a way that made you afraid that you might be physically hurt?'
>
> 'Did a parent or other adult in the household often ... Push, grab, slap, or throw something at you? or Ever hit you so hard that you had marks or were injured?'
>
> 'Did an adult or person at least 5 years older than you ever ... Touch or fondle you or have you touch their body in a sexual way? or Try to or actually have oral, anal, or vaginal sex with you?'
>
> 'Did you often feel that ... No one in your family loved you or thought you were important or special? or Your family didn't look out for each other, feel close to each other, or support each other?'

These questions don't capture the full range of experiences that constitute trauma or early childhood adversity, but they are a place to start.

You might also be wondering about where something like physical punishment lands in the realm of childhood trauma. Smacking or spanking as a method of discipline is a grey area for many parents, but it shouldn't be. In fact, the research is so conclusive that it is moving on to investigate why there is still debate. The very short answer is: don't hit your kids. A light smack on the bottom is not likely to condemn your child to a life of emotional and physical crisis, but there are far better options than pain – or threats of pain – to correct a child's behaviour. Especially because, as the longer answer demonstrates, the evidence is overwhelming that corporal punishment of children is detrimental and does not work to curb misbehaviour. A 2016 meta-analysis confirmed, based on the findings of dozens of studies, that spanking hurts and does not help kids, that it can diminish the quality of relationship between the parent and child, and can lead to aggressive behaviours in children themselves. As lead author Elizabeth Gershoff told the *Wall Street Journal*, the study was as close to a controlled experiment* as they could possibly get.

This most recent analysis was robust but so were the previous four meta-analyses, all of which came to the same conclusion: don't hit your kids. To be more specific, here were the definitions of corporal punishment used in the various studies that showed negative impact on kids in the short and long term: 'use of physical force with the intention of causing a child to experience pain but not injury for the purposes of correction or control of the child's behavior', 'a form of nonabusive or customary physical punishment by a parent or adult serving as a parent', 'physical punishment that was used primarily to backup milder disciplinary tactics'.

The persistence of both the debate and the method of punishment prompted behavioural science researchers Harriet MacMillan and Christopher Mikton in 2016 to propose strategies to move the field beyond simply trying to demonstrate that spanking is harmful. So, until a clearer strategy to stop corporal punishment is in place, we'll say it again: don't hit your kids.

In the years since the first ACE study publication, research in the area of childhood trauma has expanded exponentially, reinforcing what Felitti and his group found and more. Physical, emotional or sexual adversity all promote the excessive activation of stress and pain responses systems, changing the very nature of how those systems work. Studies confirm that compounded adversity compounds consequences – experiencing *both* fear and pain, for example, contributes to lifelong struggles. A study of former prisoners of war who were tortured found that PTSD contributes to more pain catastrophising, or mentally magnifying the intensity and fear of pain, which is related to more intense chronic pain. But some of the most

* Spanking children for an experiment is generally frowned upon by human subjects research boards.

critical findings have focused on how childhood adversity impacts later experience of pain. The chances of being diagnosed with arthritis and chronic widespread pain are higher for those who've experienced childhood adversity, regardless of what kind of stressors they may have experienced in adulthood. Longitudinal studies show exposure to adverse events before the age of seven can double the risk of chronic widespread pain at 45 years old.

We've also learned that how we experience pain in childhood has neuro-physiological implications. Those can be profound and happening at a very deep level – DNA-deep. As early as the 1930s, researchers observed that the small packets of genetic material attached to the tips of our chromosomes, called telomeres, seemed to keep chromosomes from getting attached to one another. Subsequent discoveries found these non-coding, repetitive nucleotide segments, telomeres, play a role in cellular aging. Telomeres are like the little protective, plastic caps at the end of shoelaces, as one common analogy describes, protecting the chromosome they sit on. Every time a cell divides, however, they get shorter and their ability to protect degrades. Research demonstrates that telomere shortening is the cause of age-related breakdown of cells; when the telomeres become too short, the cells can no longer reproduce, causing tissue to degenerate and die.

This is a natural part of the aging process but evidence now suggests that stress, and early adversity in particular, also works to shorten our telomeres. As if we needed any more maternal guilt, several small studies indicate that maternal stress impacts telomere length in babies – basically, the more stressed the intrauterine environment, the shorter the baby's telomeres after birth and into young adulthood. Other studies have found shortened telomeres in adults reporting high ACE scores, although this link is unclear; a 2016 meta-review of the existing studies on the relationship between psychological stress and telomere length in humans found a small but statistically significant correlation between increased stress and shortened telomeres. Right now, researchers don't have a good handle on the mechanism behind the correlation, but the point is that stress makes changes and not good ones.

The ACE study was more than definitive evidence of the correlation between ACEs and adult health outcomes; it was a call to action.

Felitti and his team pointed out the gaps in knowledge that resulted from an overly segmented approach to health, dividing physical pain from emotional, or adolescent history from presenting adult, or mental health from physiological disease. Now ACE screening is increasingly utilised in schools, homeless services, addiction services and by police departments. But integrating the ACE screening into more primary care? That's been a bit more controversial. In a 2015 interview, Felitti said that most of the push-back has been from primary care doctors who are afraid to ask the questions. Felitti described a typical response from doctors: 'My god, I can't do that, I can't ask questions like that. It's like opening Pandora's box.' Other doctors, like Dr Richard Young, a family medicine doctor in Fort Worth, Texas, don't see the utility in the screening. Young said, in the same NPR interview, 'You can't go back 40 years and make the bad childhood go away.' Doctors' reactions reveal the drawbacks of the biomedical framework – if they can't fix it, they don't want to hear about it.

This perspective not only dismisses the opportunity for preventive intervention and strategies based on a more complete picture of the patient, but it also, according to Felitti, ignores the healing potential that comes from basic self-disclosure, from sharing pain. Simply telling another person about our most painful experiences, which are often kept in secret out of fear or shame, can do wonders. As Felitti said, 'They leave with the understanding that they're still an acceptable human being, still part of a group.' He thinks this, at least in part, explains his findings that among more than 100,000 patients, those who were asked the ACE questions went to the doctor's office 35 per cent less often than those who did not. At the same time, there is no evidence that asking patients about their history *is* opening Pandora's box. Interviews with patients who have undergone the ACE screening suggest they do not have an expectation that their primary care doctor is going to 'fix' anything, nor do they become inconsolable.

We're still plastic

This might be a good spot to stop and offer some points of balance and, we think, hope. About two-thirds of the original sample Felitti surveyed experienced at least one ACE, and about 33 per cent of children and about 80 per cent of adults in the UK have experienced or will experience some sort of traumatic event. This means most people you meet will have experienced at least one. But the lifetime prevalence for PTSD is only about 5 per cent among adolescents and 6.8 per cent among adults in the US, and about 1 per cent of kids and 4.5 per cent of adults in the UK. This means that, as a species, we're pretty damn resilient.

There's an inspirational quote that 85 per cent of people on Facebook have already or will at some point pin to their page*: 'Be kind. Everyone you meet is fighting a hard battle that you know nothing about.' A quick investigation into the origins of the quote doesn't really yield much – Plato, Aristotle, a semi-famous mid-century Scotsman, thriller writer Brad Meltzer. But perhaps the fact that it doesn't seem to have a verifiable origin underscores its intuitive, organic quality. It feels accurate because we all can feel like we have secret battles we don't share, that cause us pain. Acknowledging that dangerous adversity is widespread and does real harm can go a long way in fostering empathy and in diminishing the effects this kind of adversity has on us. Resilience can emerge from a terrible trauma, as the lived experience of countless ordinary people demonstrates.†

Just as biology, cultural context and genetics are not destiny, neither is a high ACE score. We all know this. For example, according to NPR's ACE screening questionnaire, Linda has an ACE score of 7 on a scale of 0 to 10. This is high; in the original ACE study, only 12.5 per cent of those surveys had an ACE score of 4 or higher. At a score of 4, the risk of developing life-threatening, chronic illnesses

---

* We made that figure up.

† Another bright spot: ACEs are tentatively linked to more empathy. According to a 2018 study from City University of New York and University of Cambridge 'Results across samples and measures showed that, on average, adults who reported experiencing a traumatic event in childhood had elevated empathy levels compared to adults who did not experience a traumatic event. Further, the severity of the trauma correlated positively with various components of empathy. These findings suggest that the experience of a childhood trauma increases a person's ability to take the perspective of another and to understand their mental and emotional states, and that this impact is long-standing.'

increases substantially: for chronic pulmonary lung disease, it's as much as 390 per cent; for depression, as much as 460 per cent. But Linda is healthy, with no history of depression, no drug or alcohol addiction, and a resting heart rate of 57bpm. She doesn't smoke, has been in a stable marriage of more than 10 years, and doesn't spank her kids or pets.

Linda's childhood and adolescence were unfortunate in some ways, but the weight of violence, fear and unpredictability was balanced by several important elements. Research demonstrates that even one good relationship can mitigate ACE effects; having responsive relationships generates healthy brain structure so that when toxic stress does come it can be managed, and setting in place structures that reinforce core life skills can help cushion the effects. For Linda, her mother was a strong, unwavering source of support, and her competitive school environment inculcated core life skills (although it didn't help with her tendency to procrastinate). Ultimately, adverse childhood experiences that sank other members of Linda's extended family were mitigated in her life by her peculiar circumstances. But crucially, transforming adversity into resilience isn't some alchemical process that's the duty of the individual alone – all of the research underscores the importance of communal support, institutional and individual empathy, and structure.

The other good news is that we're not locked into ways of dealing with pain. Despite the fact that we've just spent a whole chapter telling you how desperately important the first few years of our lives are in shaping how we respond to pain and stress in later life, there is always room for growth. The circuits and responses that make our experience of pain, and how we respond to it, are not 'hardwired' – our brains are miraculously neuroplastic, in ways that we're only just now learning.

You may have heard the term 'neuroplasticity', referring to the ability of our nervous system to adapt and change according to the environment, experience and demand. For decades, science held that after childhood, it was no good trying to teach an old dog (or even just young adult) new tricks – we could neither create new neurons nor change the function of our existing neurons, and all us old people could do was stand by as they died off in their millions. But our nervous systems are designed to be adaptive and plastic. Though they are not as flexible as during childhood, studies show that our brains continue to exhibit neuroplasticity indefinitely and that neurogenesis – the growth of new nerve cells – is a lifelong process.

That neuroplasticity is built into our systems is important in understanding how we feel and experience pain, how we will do so throughout our lives and how, potentially, we can have more control over that experience. If our nervous systems can form habits, then we need to make sure we're forming the right ones; if we've formed bad ones, then the hope is that we can perhaps work to replace them with good ones.

Acrobat instructor Zach Fischer doesn't only work with children and teens – he also teaches adults, many of whom are new to the art of throwing their bodies around a padded room. But in truth, teaching them how to vault over a park bench is only part of his work. The other part is reminding them there are always opportunities to learn. 'You are still plastic even as you age,' he says. 'So my job is a lot to encourage plasticity, encourage the idea that you can change yourself if you work slowly and repeatedly at it.'

Getting adults to believe they can have new emotional and physical experiences can be hard, even those adults who have signed up for Fischer's classes; pain, for most of them, is bad and to be avoided at all costs. 'Adults have had a long time to convince themselves this is who they are, and this is all they will ever be and there is no possibility for them to be something else,' says Fischer. He has to remind adults of the joy and power that can come through fully inhabiting their body. After adolescence, after we stop throwing ourselves down hills just to see how it feels or spinning in circles to fall on the ground and delight in the dancing clouds above, it can be easy to forget that our body is central to how we learn and is full of opportunities for sensational experiences. Our body stops being both a vehicle for and the site of adventurous exploration and becomes the thing to be protected, guarded.

A big challenge is getting adults to, as Fischer says, 'explode at their maximum power'. During childhood, doing anything as hard or as fast as you can was part of the point, just to see what it felt like to go full throttle. As adults, however, we've already learned the consequences – our very efficient nociceptive system has made sure of that – and we pull back out of fear of damage or pain. 'Adults are often so averse to intense sensation that getting people to even feel like they're falling, to just feel the experience of letting gravity take you – and you have to do something in order to not faceplant – it takes work and time,' explained Fischer.

Part of that work and time *is* getting hurt – and relearning what that means. As Fischer points out, pain and injury in Western culture is frequently perceived as evidence we've done something wrong (or have been done wrong). But in his gym, pain and injury don't mean that; rather, they mean learning and growth and discovering that our bodies are not as delicate as we might think. The pain of sore muscles or a failed trick can be good to feel and, in identifying and naming those instances, we expand our emotional diversity and granularity. Fischer helps his adult students practise this in the same way he helps his younger ones: by actively reframing and renaming what they're feeling in the moment in a useful way. Just as he asks his younger students, Fischer asks adults, 'What is the sensation beyond the pain? Is it pressure, sharpness, internal tearing, soreness? We try to delineate the different types or flavours of pain in a way and gain perspective on the intensity or amplitude.' At each step is the opportunity to experience a sense of agency or control, and to establish the self at the centre of defining the experience, not the pain.

This exercise works even when you're not trying (and failing) to land a backflip. Having a greater awareness of our emotional states enables us to both perceive the good things in life more and manage the tough things better.

As we learned from Bastian in *The NeverEnding Story*, the most emotionally compelling and heart-wrenching movie for those of us born around 1980,* naming is powerful. Words give us something to hold on to, a scaffold on which we can organise diffuse, overwhelming and confusing feelings. This is especially critical for when we feel bad. Whether it's acute or chronic, physical or emotional, pain can be silencing and isolating; as Elaine Scarry

* Artax in the Swamps of Sadness. We're not crying, you're crying.

wrote in her seminal book, *The Body in Pain*, pain 'destroys language'. But by naming it, and shouting it out the window like Bastian, we wrest back control and can create something new. Just as it works in helping Fischer's adult students reconnect and re-experience physical sensation, it also helps when the pain is deeper, buried in our memories: As Bessle Van der Kolk wrote in his highly influential book *The Body Keeps the Score: Brain, Mind, and Body in the Healing of Trauma*, 'Discovering yourself in language is always an epiphany, even if finding the words to describe your inner reality can be an agonizing process.'

We need (some kinds of) pain

Kids need pain – and so do all of us. We all need it for the same reasons: not only to help us make better decisions next time but also to show us that we can handle stress and discomfort, that we are resilient and brave, that we will get through this. And we need to be able to express the flavours of pain, to discuss and describe pain without falling back on words like "ouch". We need to be able to give ourselves some measure of control over an experience that often feels anything but controllable. This message is crucially important now. The natural consequence of living in a society inundated with messages about safety is that we are more afraid. Being afraid of pain means that we not only avoid pain but also any situations that might carry the risk of pain. But situations that carry the risk of pain also promise other possibilities, such as joy, excitement, satisfaction.

Pain adds to our emotional vocabulary, not only by itself being an endlessly variable emotion but also by making sure that we are likely to experience other emotions too.

Rebuilding our relationship with pain and reconnecting to our bodies in adulthood requires an act of courage. Before we can name and reclaim, we have to step off the end of the diving board and *do the thing* – as Fischer says, explode at our maximum capacity. This step is the hardest part for a lot of people, and it's often not about the pain but about the fear of pain.

Fischer's adult students often say to him, 'I don't know what will happen.' 'What will happen,' he tells them, 'is you'll survive.' But as we've pointed out in this chapter, knowing something logically is not enough. We're about to meet some people who can teach us how to believe it.

CHAPTER FIVE

A Visit to the Pain Cave

Exactly how Adharanand Finn found himself hallucinating in the freezing dark, legs in agony, running down the side of a mountain in the French Alps is a story with many roots. Maybe it starts when he was a primary school kid with a passion for running. Or maybe in his twenties, when he won a 10K race after years of indifferent training squashed in between life stuff. Or maybe it was on the 19-mile fell race in the Scottish Highlands, chasing his equally competitive younger brother. Or the sand dunes of Oman, where he ran 100 miles over six days. Wherever his story started, it was quite possible that it was going to end here, with a stumble, a scream and the scrape of falling rocks.

It was into the unplanned second night of the Ultra Trail du Mont Blanc – better known as the UTMB, covering 171 kilometres through the French Alps and climbing nearly 10,000 metres – and Finn was seeing buildings, tents, aid stations, *people* who weren't there. His quads weren't working and he hadn't slept in more than 30 hours. He nearly walked off the side of the mountain. He nearly gave up. This, he realised, was the pain cave, the dark night of the soul that ultra-runners talk about.

Finn spotted another runner and decided, he said, to fill his field of vision with this runner's back. 'When we started going down, this person was going down way quicker than I was comfortable with, but I thought, "I just need to stick with this person",' he told us. 'So my legs felt completely destroyed, and I was struggling with each step, but as I had to go quicker, I became more and more comfortable moving at this speed. And all the sensation of tiredness went out of my body, and then I started to have this sensation of warmth.' Finn managed to make it off the mountain, finishing the race in 43 hours, nine minutes and seven seconds. 'Weirdly, the person never spoke to me,' he recalled. 'I'm pretty sure it was a real person…'

Finn, a journalist for the *Guardian*, has been a runner for most of his life. In 2011, he spent six months training on dirt trails with a running club in Kenya, a country that seems to produce more elite distance runners than any other. His book *Running with the Kenyans*, detailing the country's running culture and his efforts to slot himself into it, ends with his sub-three-hour performance in the New York Marathon. Even so, when a magazine editor asked if he'd be interested in running 102 miles over six days through the desert in Oman, he said he wouldn't. And then, he started thinking about it.

'There was an attraction to finding out what it was like to run that far,' he said. 'When I thought about it more from the point of view of a life experience – this one was in the desert, 100 miles in Oman – the idea that I could traverse the desert on my own, carrying all I needed, it could be an outlier experience, an opportunity that not many people could get to do.'

So he said yes.

Since at least the dawn of agriculture and permanent human settlements, people have engaged in sport. Wrestling, running, swimming, variations on throwing heavy or pointy objects, archery – mostly military exercise disguised as sport, but sport nevertheless. In the 19th and 20th centuries, sport took on increasingly organised, modern dimensions. In the UK and the US, spectator crazes for sports like 'pedestrianism' – fast-walking contests – morphed into formalised sporting events. Leisure time and energy were channelled into football and cricket clubs, running and rowing clubs, cycling and fencing, baseball and basketball teams. By the time of the first modern Olympics in 1896, small-scale sporting festivals held across Europe and the US had long been popular tests of athletic prowess. With a solid proof of concept, these activities, including a variety of foot races, were growing into something much more marketable and widespread.

More recently, the interest in endurance athletics has grown exponentially. USA Triathlon, which represents Team USA in the Olympics, reported a surge in membership from 127,824 in 1999 to 510,859 in 2012. Obstacle-course racing, the kind that pits participants against muddy bogs and A-frames on courses of varying lengths, has also exploded in popularity: in the US alone, the number of runners rose from nearly 70,000 in 2010 to more than 500,000 in 2015, according to RunRepeat. The Spartan Race, one of the biggest names in the current industry, claims to have more than 1.3 million participants worldwide annually.

Ten to 15 years ago, ultra-running was largely a few brave souls running in a handful of impossible-sounding races.* Since then, however, the extreme sport space has exploded, fuelled in part by social media, particularly Strava and Instagram – if a runner bounds through the forest and no one sees her, did it really happen? According to *Ultrarunning Magazine*, the number of people who finished an ultra in 1980 was 629. In 2016, it was 88,075. That's not a huge amount – 53,520 people finished the New York City Marathon in 2019, more than 40,000 ran London – but it represents an increase of nearly 14,000 per cent.

An ultra-marathon is any race that pushes past the 26.2-mile mark, already a huge challenge involving cramps and soreness, intestinal distress, bleeding nipples and blackened toenails and so much chafing. By the time Finn ran the UTMB, he'd already run seven ultras – he certainly knew what he was getting into. And it's not pretty. Describing the further toll that running past that point takes on the body can devolve into a bit of voyeuristic pain porn, but it's worth examining to get an idea of how much pain these athletes are willing – and able – to endure.

It can take an average of two weeks to return to baseline stress and recovery levels after a race, which seems rather short given all the trauma that happens to your body as you're pushing it so far beyond what it wants to do. For starters, the repeated impact of your feet on the ground is, to put it mildly, jarring – the average ultra-runner lands

* Ultra-marathon has its roots in the early 19th century, with the history-making walk of Captain Robert Barclay Allardice. Barclay, then a 29-year-old gentleman, walked 1,000 miles in 1,000 consecutive hours, earning him huge amounts of money in wagers. Others tried to emulate his good fortune; in 1815, for example, thousands of people descended on Blackheath, then grasslands just outside of Greenwich, to watch George Wilson walk 1,000 miles on a pre-measured course around the common. Pedestrianism bouts were routinely organised by pubs, and the sport soon made its way to the US, even Madison Square Garden. The kinds of injuries these brave athlete-showmen suffered were very much the same as in modern ultra-runners: cramping, sleep deprivation, outraged joints and chafing. Lots and lots of chafing.

on his or her feet 5,000 times an hour during a race. Stress fractures in the feet, the leg bones and pelvis are common. Marathon runners experience this, but the incidence of 'running-related musculoskeletal injuries' increases with ultra-runners, and the most common – Achilles tendinopathy and patellofemoral syndrome (runner's knee) – are found in up to 18 and 16 per cent of ultra-runners, respectively. Feet and legs swell, although research puts this down to the large amounts of fluids ultra-runners take on during a race. In fact, though they definitely need fluids, a tendency to over-hydrate can lead to exercise-associated hyponatremia, a potentially fatal complication in which the levels of sodium in the blood are dangerously diluted to the point of impairing organ function.

Gastrointestinal problems are the number one reason people don't finish races. As you run, your body starts shifting its blood supply towards the working muscles and away from the gut. Food is digested far more slowly, meaning that it just sits in the stomach, a ticking time-bomb of gastrointestinal distress. Cramps, nausea, vomiting and diarrhoea are all likely and all symptoms that increase with distance. And speaking of cramps, your stomach isn't the only place to suffer. The exact mechanism behind cramping isn't entirely clear – it could be a kind of misfiring of the motor nerves charged with managing muscle contractions – but exhaustion and electrolyte depletion likely play a role. Whatever the reason, more time spent using your muscles means that legs, shoulders, back and pretty much everywhere is a potential cramp waiting to seize up.

Then there are the course conditions themselves, which, depending on where you're running, can induce hyperthermia or hypothermia. Wind can whisk away the protective layer of liquid over your cornea so that the cornea swells, causing blurred vision and, in some cases, temporary blindness. Uncertain terrain leads to falls, cuts and bruises. Ultra-runners also exhibit higher rates of respiratory ailments, quite possibly due to dusty trails and exciting flora. We haven't gotten to the chafing, blisters or, for the races that last more than 24 hours, sleep deprivation – the hallucinations that Finn experienced are pretty common and are typically a result of not having slept.

But the pain doesn't stop when they stop running. Recent research shows just how deeply this type of sport changes how we process pain, potentially via changing the firing threshold – adequate stimulus – of various somatosensory neurons. Athletes, for example, are more likely to suffer from chronic widespread pain. The reasons behind this are not yet well understood. Until recently, most research on athletes focused on how they adjust their perception of pain during physical challenges. But a 2013 study found that though athletes have a higher pain tolerance, their pain inhibitory systems – the ones responsible for kicking up the circulation of endogenous opioids – are less responsive when compared to non-athletes.

This is likely due to the frequent and long-term activation of what researchers call the Conditioned Pain Modulation (CPM) system. The CPM is what allows athletes to endure 100-mile marathons, through the judicious release of endogenous painkillers. But by engaging in these extreme endurance challenges routinely, their bodies have essentially recalibrated their pain inhibitory systems to a higher threshold. Researchers describe this as a 'ceiling effect', meaning that an athlete's body does not activate its CPM and therefore doesn't deploy endogenous painkillers until the pain reaches a higher intensity. In other words, their bodies respond to everyday aches and pains with: 'You think that's bad? That's not bad at all. Call me when you're running another marathon.' They might feel the pain more, because they don't get the endogenous help that other people who don't do extreme sport get. (Says Finn, 'I'm one of those people who when I stub my toe, I make as much noise as possible.')

Athletes also experience flavours of pain differently to non-athletes. For example, researchers found that athletes have a higher threshold for experiencing mechanical pain, as measured by a pinprick test, but a lower threshold for experiencing vibration. This could be adaptive: a lower threshold for vibration could be the result of athletes developing well-trained locomotive systems – basically great agility and balance – that allow them to adjust their movement based on feedback, like vibration, from the environment. Taken together, the changes observed among athletes points to just how adaptive and active our nervous system is, constantly recalibrating based on our changing needs.

Ultra-runners talk a lot about pain – specifically, about 'the pain cave', Finn says. 'Most ultra runners will know instantly what you mean. It's not the same kind of pain you get if you've broken your arm … it's more debilitation, you feel completely debilitated, every ounce of your body is pleading with you to stop.' And the worst part is that you *can* stop. At any time. 'There is just this huge temptation to do what your body is screaming at you to do, and to carry on in that situation where your quads, your ankles, your neck, your arms even, everything is aching, kind of weeping at you – you have to make this conscious decision to carry on.'

Children need painful experiences to learn how to respond to pain, but learning doesn't stop once we reach adulthood – that's the neuroplasticity we're talking about. Sport offers adults and teenagers, much like risky play for children, an opportunity to experience and, critically, to manage pain. Because pain is part of the process, and because injury is likely, it gives them an opportunity to find themselves resilient. And that's certainly part of the attraction.

When we talked to Finn, the UTMB had been eight months before and he hadn't done another ultra since. 'I had this thought, this moment in time, a very real thought that I'd rather chop my hand off than do that race again,' he said. 'I struggled badly.' This was after he'd finished and was relaxing with his family in the runners' campsite – although he was still hallucinating, seeing charming French cows who weren't actually there in the charming French countryside. But now, he says, the idea of doing another one is really appealing. As he wrote in his 2019 book chronicling his experiences, *The Rise of the Ultra Runners*, 'The memory of the pain dissolves so quickly.'

It's easy to think that people like Finn are a different species, that they have something the rest of us – who are, for the most part, not ultra athletes – do not. But endurance is not a superhero skill, they're not born with the ability to manipulate their physical response to pain. They learned it.

How we endure

In some ways, being an athlete is improbable. As Harvard evolutionary biologist Daniel Lieberman and others suggest, for millennia, human evolution has favoured energy efficiency. Allostasis – the intrinsic process of regulating things like blood glucose levels, blood pH, and body temperature – is a substantial task most of the time, especially with our big brains consuming as much as 20 per cent of our energy production. Exercise, especially the extreme kind, tends to upset that balance, burning through glucose, messing with our temperature, dehydrating us and making allostasis that much harder. This inclination to preserve energy is also why, when faced with the choice between bingeing on Netflix and running a marathon, many people will find a comfy spot on the couch.

Except some of us *do* run marathons – and longer – and we've always wanted to know how the human body endures the demands we put on it. From the 19th century on, science had new tools to investigate the answers, including physical chemistry and an understanding of how blood circulation and musculature work. In the 1870s and 1880s, Italian physiologist Angelo Mosso* theorised that fatigue was the by-product of chemical processes that involved the generation of carbonic acid, among other toxic substances. Building on the toxic chemical by-product theory, in 1922 Archibald Hill, then a professor of physiology at the University of Manchester, shared the Nobel Prize with Otto Meyerhof for their work on muscle glycolitic metabolism, investigating the chemical and mechanical circumstances surrounding muscle contraction. Meyerhof famously stuck a severed frog leg in an airtight jar and forced it to contract using electrical impulses. The breakdown of muscle glycogen in the contraction produced lactic acid; after too many zaps, the leg stopped contracting, leading researchers to conclude that it was the lactic acid

* Mosso was also quite possibly the first person to try to understand the workings of the brain using non-invasive neuroimaging. In 1882, he built a device to measure brain activity, which he called the 'human circulation balance'. Mosso's theory was that the harder the brain worked, the more blood it needed. The device was essentially a wooden plank balanced on a fulcrum; subjects were asked to lie down on the plank, their balance calibrated, and then were tasked with thinking exercises. Mosso found, just as he hypothesised, that the more strenuous the intellectual activity, the more blood the brain needed, tipping the balance of the scales. In 2014, researchers at the University of Reading built a human circulation balance based on Mosso's specifications to test his findings – and it worked.

that inhibited the leg's ability to contract. Hence, muscle fatigue arises from too much lactic acid.

More recent research demonstrates that this is a somewhat oversimplified explanation. A study from 2014 did find that injecting subjects with a specific combination of metabolites – chemical by-products of intense exercise – produced the pain and fatigue associated with exercise, without the exercise. According to the University of Utah researchers, a cocktail of lactate, protons and adenosine triphosphate (a form of cellular fuel) made subjects feel as if they'd just run a marathon; this, they suggested, triggered activation of receptors implicated in nociceptive pain processing. But presence of these chemical byproducts alone does not necessarily bring your body to an exhausted halt; we also have to get the pain signals activated by those byproducts. In 2008, those same University of Utah researchers decided to see what would happen if athletes couldn't receive messages of pain or fatigue from their somatosensory system. In a double-blind, placebo-controlled experiment, they dosed up cyclists with fentanyl, effectively cutting off the signal from their peripheral locomotor muscles, then asked them to cycle as fast as they could on a stationary bike for 5 kilometres. They felt great during the exercise, but they couldn't walk afterwards – they couldn't even unclip their feet from the pedals.

Outside of the lab, our home-grown pain inhibitory system can make us feel pretty great during a run, or can at least counteract the feeling of imminent collapse. This is called 'acute exercise induced analgesia' – more often referred to as 'runner's high'. Clinically, the runner's high is characterised by four components: anxiolysis, or a decrease in anxiety; analgesia, or pain inhibition; sedation, which may or may not be a positive during a run; and euphoria. Other common descriptions include a lost sense of time and a feeling of relaxation, pleasantness and effortlessness. And it is a real thing, although the degree to which it impacts mood or is experienced as a 'high' varies widely among runners.

The dominant theory explaining the runner's high has, for a long time, been the 'endorphin hypothesis' – basically, that running elevated the circulation of our endogenous opioids, which made runners feel good and in less pain. In a 2008 *Cerebral Cortex* article, a group of researchers from Germany tested opioid binding of 10 long-distance runners while at rest and then again two hours into an endurance run. They found that, at two hours in, self-reports of euphoria had significantly increased and receptor availability for opioid binding had decreased, supporting the opioid theory of the runner's high; moreover, they noted, 'endogenous release of central opioids occurs preferentially in brain regions belonging to frontolimbic circuits that are known to play a key role in emotional processing'. However, it is unlikely that any single neurochemical is working independently in the runner's high, as highlighted in a recent study out of the Penn State College of Medicine. In the study, saliva of 25 runners was analysed and the results found significant changes in endorphin, endocannabinoid and GABAergic, which refers to the control of neurotransmitter GABA, signalling – all of which are related to reducing anxiety and pain – among those reporting a runner's high.

But physiological accounts of pain in sport and our body's responses to it are not enough to explain why – and how – exhausted, aching runners keep running. Because in every race he ran, there came a point that Finn thought he would stop, had to stop, no amount of endogenous opioids was enough. But he didn't. What kept him going was his conviction that the pain, the debilitation, was all in his mind. In one ultra-race, Finn had just run down a steep hill and his knee started aching. It was only two hours into the run and he wasn't ready to call it quits. 'I laughed at myself, and I thought, "You're going to have to do better than that!"' he said. His knee stopped aching. 'I was just trying this, trying to talk to myself and think, "No, that's not a real pain." I thought that it wasn't going to work, but it did.'*

Finn's cognitive coping mechanism has worked in every race he's run so far. There came a point in every race, often fairly far in, he said, when the pain would just *go*. The aching quads and screaming Achilles tendons, the burning sensation in his feet – it would all just

* Finn said at one point, he tried to 'do it for his kids', to use them as his mental motivation. 'Then I realised my children didn't care.' Children are notoriously indifferent to their parents' pursuits. True story: Linda runs on a cross-country team and had just stumbled over the finish line after a cold, rainy, muddy race up several steep hills. Her oldest son, then eight, looked up from his iPad and yelled, 'What took you so long?'

disappear. 'I came to the conclusion that a lot of the pain that I was experiencing earlier on in the race was imagined. My brain was trying to slow me down by telling me that everything was aching.'

This isn't a straightforward 'mind over matter' argument. Finn *was* undoubtedly aided by the release of his endogenous painkillers and by his brain and body's modified pain threshold as a result of all of his training. And stress fractures, cardiac arrest as a result of exhaustion, and dehydration are all real things; the pain and fatigue we experience before those occur are signals that we may be overreaching, so it's important not to just power through blindly. However, cognitive coping strategies are what enable us to be a bit more sceptical of the signals we're getting.

Finn's experience tallies with a 2007 study of Olympic cyclists by sports behaviour researchers Jeffrey Kress and Traci Statler at California State University, looking at how the athletes coped with exertion and pain during performance. Kress and Statler found it wasn't only acute exercise-induced analgesia that insulates the cyclists from pain but also the cognitive strategies they employ. The most used strategy? Reminders that 'pain was purely perception'. That, combined with positive self-talk, goal-setting, reframing the pain as a positive thing, reappraisal (such as telling themselves the pain feels like vibration and is soothing) and viewing pain as useful, part of the sport and identity are all frequently cited cognitive coping strategies among endurance athletes.

From the perspective of the constructed emotion, these strategies make sense: recategorising your sensations as something other than pain can help you push past that pain. 'Whenever you exercise just until you feel unpleasant and then stop, you're categorizing your physical sensations as exhaustion,' wrote Lisa Feldman Barrett in *How Emotions Are Made*. 'You'll always exercise below your threshold, despite the health benefits of continuing.' Choosing to recategorise the unpleasant sensations as something else – even, as Jane Fonda would say, to 'feel the burn', a positive yet physically taxing affective state – you can push harder and longer. In fact, a study found that public speakers who recategorise their anxiety as excitement performed better; though their sympathetic nervous system activity still increased, analysis found lower levels of the proinflammatory cytokines that tax the system and lower mood. Just as each strike of the foot on pavement fires off nociceptive signals hell-bent for the brain, so can our own thoughts work to generate neurophysiological changes that allow us to carry on.

Cognitive strategies also imply a degree of acceptance, of what people in the military, triathletes and endurance athletes term as 'embracing the suck'. The effectiveness of 'embracing the suck' was demonstrated in a 2018 study published in the journal *Pain* investigating what psychological factors that may explain the higher pain tolerance were observed among endurance athletes. Researcher Gregory Roebuck compared 20 ultra-runners to 20 age- and gender-matched controls. Supporting previous findings, the 20 ultra-runners were able to hold their hand in super-cold water significantly longer compared to controls. The biggest factors explaining the variance: ultra-runners are less anxious about pain and reported less escape or avoidance tendencies. Basically, they weren't afraid of pain.

What Finn and athletes like him have learned is that pain is not something that just happens to them; they have a say in what it means, in how it is experienced in the present and remembered in the future. Athletes don't just ignore the pain and push past it. They assess it and reframe it. They've learned how to negotiate with pain. They've learned to hear and then sometimes politely ignore the body's alarms. They take control.

Choosing how to experience pain

Control is one of the most significant influences on our experience of pain – just being able to choose to close your eyes can make a difference in your perception of pain intensity.

Having some control over what's happening to you impacts both how pain is felt and pain-related neural activity. Researcher Dr Catherine Hartley illustrated this point in a study she presented at the annual Society for Affective Science conference in Boston in 2017. In the experiment, participants played a computer game in which they received random electric shocks as an on-screen dot moved through a maze to a 'safe' zone, where the shocks stopped. Half the subjects were given control over the dot's movement and could navigate it to the 'safe' zone. The other participants could only watch the dot move through the maze. The passive subjects, those just watching, reported feeling more pain compared to the players who had control. Fascinatingly, this relationship held up even when there was no 'safe zone' to navigate to that would stop the shocks. Just being able to control the movement of the dot through the maze made a significant difference in subjects' pain ratings.

Agency is intimately related to control, but they are not the same thing. Agency is the *perception* of how much control we feel over actions and consequences. Lack of agency is a hallmark of some of our most painful experiences – this is the feeling that Jude Taylor described in Chapter 2 when she related being stuck in the hospital, 'trapped in pain', wanting so badly to be able to do something to relieve her pain but not being able to. Our sense of agency is measurable; one way is by measuring people's perception of time. When we are the one to initiate an action, the perceived time to outcome is shorter, regardless of how long it actually took. Here's an example: pressing play on your remote to start your favourite streaming show. The show is going to start when your internet TV decides it's ready to start, but it seems to feel like it starts faster when you, instead of your partner, press the button. This distinction is important because, while we'll all face painful instances where we may not have control, such as childbirth, we can still find opportunities to increase our sense of agency.

Preserved agency in the face of threat is often the difference between an experience being scarring or empowering. And learning how to gain a sense of control over an experience – pain – that seems predicated on *not* being in control is what athletes do. But how can you maintain a sense of agency when you're being punched in the kidneys? How can you reframe *that* kind of experience?

For the better part of the last 25 years, Rami Ibrahim has spent his days and nights getting punched in the face, kneed in the ribs and kicked in the shins. Ibrahim, also known as 'The Arabian Nightmare', 'Cinderella Man' and 'Son of Palestine', is the longest-active Muay Thai fighter in the US. Muay Thai (or 'Thai boxing', also called the 'Art of Eight Limbs') is a martial art that combines boxing, kickboxing and grappling. In Muay Thai, every edge of the fighter's body is in play, from the elbows to the shins, fists to the knees. It is a brutal and effective form of self-defence with roots in 18th-century Thailand,

and Ibrahim is really good at it. At 17, he became the youngest American ever to win the US Muay Thai championship. He's won the World Boxing Council Muay Thai International Championship, the World Kickboxing Association United States Muay Thai Championship, the United States Kickboxing Association International Muay Thai Championship, among many others. In May 2018, just before his retirement after 186 bouts, he became the first Palestinian American to win a unified World Kickboxing Association/United States Kickboxing Association Kickboxing World title.

'This is a serious sport where the objective is not to score points on your opponent,' explained Ibrahim. 'Let's be real: the objective is to hurt your opponent.' Ibrahim knew that every time he got into the ring, he was going to be hurt; it was just a matter of how seriously. Now 37 and the father of a young daughter, Ibrahim no longer fights professionally. Instead, he coaches junior fighters at the gym he founded in 2007, Rami Elite, less than a mile from the Philadelphia neighbourhood he grew up in. At the entrance is a life-size photo of Ibrahim against a red background, outstretched arms bearing the weight of eight massive championship belts, another around his waist; the expression on his face is a mixture of humility and pride, the face of someone who believes that both Allah *and* he got him where he is today.

Ibrahim started fighting in the early 1990s. He was born the youngest of six in Kuwait – where his Palestinian parents settled after the expulsion of Arabs at the formation of the Israeli state – but when Iraq invaded in 1990 his family were forced to flee. In some ways, however, they just left one war zone for another. In north Philadelphia, crime was high; once, he said, a fight between rival gangs left blood on his doorstep. Plus, he was a 10-year-old Arab refugee who didn't speak English. 'It was very, very bad. It was very bad. Very painful emotionally. I was in a very bad neighbourhood,' he told us. 'I used to get into fights all the time, all the time. I used to get rolled on, jumped by kids 'cause I was the odd one.'

Ibrahim's parents signed him up to a martial arts gym, both as a way to keep him busy and to teach him how to defend himself. He had talent and he had drive. What he didn't have were mentors or, it seems, an off-switch. 'My trainers didn't know too much, but they told me, you know, "You need to be a killer, you need to go out there and win at whatever cost,"' he said. 'So they brought me up to think like a young lion … I was just crazy. I would go all out, you can just see it in my face. You know, I wanted this win at any cost.' Even if that cost was himself.

When he was 16, Ibrahim's coach matched him with a fighter who turned out to be a professional from overseas, 10 years his senior, stronger and more experienced. This fight was technically illegal, as Ibrahim was an amateur, but at the time there was no strict oversight from a governing commission. 'If it wasn't for my heart, I would have been dropped in the first round. But unfortunately – and I say unfortunately because I wish I had gotten knocked down or took a knee or something – I finished the whole fight without getting stopped and I remember that fight more than anything,' he recalled. This fight *hurt* and it is the fight he describes as 'the first time I got a beating.'

His coach should have stopped the fight, but he didn't. 'I remember my coach was trying to get me to 'wake up',' Ibrahim said hesitantly, struggling to find the words. 'That was wrong of him. You know, he brought my dad to the corner to try to get me, to motivate me. That's one of the most painful things I remember, and my dad, he said, "Come on, son, do it for me." I really wanted to do it for him. But I couldn't … And that's the first physical painful memory I had.'

Ibrahim's coach wanted him to be a 'young lion', a 'killer'; he pushed and manipulated him through a bout that he shouldn't have been fighting in the first place. More than that: Ibrahim was the one in the ring, throwing the punches, but he wasn't in control, he felt controlled. And though he's been injured dozens, hundreds of times, this was the time that hurt.

A study published in 2018 by a team in Belgium offers insight into Ibrahim's experience. It is not a matter of having agency or not, but rather that agency exists on a continuum and is influenced by social context. Using perception of time as a measure of agency, researchers found that participants assigned to act as an 'agent' to carry out orders of a coercive authority figure, the 'commander', perceived the time between their action (pushing a button) and the outcome (a ringing tone) as longer. Though the agent was still technically in control, they felt like they weren't.* Put simply, carrying out someone else's orders, doing it because someone else tells you to, reduces a sense of agency, of control over your actions and their outcomes. And as Ibrahim learned, less control equals more physical and emotional pain.

Over the years, Ibrahim learned to focus inwards, to listen to his body so that he was the one deciding when pain and injury were acceptable consequences. And they often were, as pain was no match for his passion. When Ibrahim was 23, he broke his fibula in a boxing match. It was in the second round; he knew right away when it happened. 'I was trying to throw punches, I could hear, I kid you not, the bone go [creeaak] and I felt it, I hear it through my ears,' he recalled. He finished the round, but told his coach that he couldn't feel the ground any more. This time, his coach wanted to stop the fight and it was Ibrahim who refused: 'Coach, give me one more round, and if I do bad, you can stop the fight.' His coach agreed and Ibrahim got back up.† 'So I just sucked it up and I tried my best not to show any pain, just so my coach don't stop the fight. And I finished that entire fight,' he said with a smile. He couldn't walk for two months. 'I remember, my mom was so happy. I'm like, "You're a crazy woman, why would you be so happy? For something like this?" And she said "Because now you'll stop fighting." And I said "Nah."'

Predicting the future

Control is integral to how Ibrahim experiences pain, and whether that pain is scarring or empowering. But he wins, he says, because he goes into the ring prepared for the pain. He knows how much it's going to hurt when he delivers a blow to his opponent's head, he has imagined his opponent's knee connecting with his rib cage as he pivots around him. Ibrahim knows this because he has experienced it

* Interestingly, the commander also experienced less agency when ordering the agent to carry out their order.

† You know who else got back up? Rocky. You know who else is from Philadelphia? Rocky.

in the past, can remember and imagine it in the present and strategically plan for it in the future. 'I tell myself, "What's the worst thing that can happen to me?"' he explained. 'So I try to face the most fearful thing. And then you know, as long as I can acknowledge it, face it, face fear head on, then it can't shock me any more ... 'Cause that's what hurts us, it's the element of surprise.'

Even as Ibrahim does the cognitive work of thinking through the worst-case scenarios, setting his expectations of pain appropriately to help him manage it when it does come, his nervous system is also preparing. And once he's in the ring, it's adapting in ways that he is barely aware of.

Even before Ibrahim throws his first punch, changes in neurochemical signalling are happening – ramping up sympathetic nervous system activity, increasing the circulation of endogenous opioids – that will buffer the blow when his fist lands. This predictive ability is possible thanks to allostasis, the perpetual process of internal regulation, calibration and balancing energy in and energy out. Allostasis, in turn, is able to regulate efficiently by predicting our interoceptive processes, the kind of real-time 'here's what's going on' network, composed of afferent sensory input from or information about peripheral organs, tissues and physiological processes.

Breaking down the neurophysiological mechanics of the predictive allostatic-interoceptive system can get a bit complex, so we'll start with a metaphor: in business, a good manager is one who identifies, reacts, and solves problems quickly. But an excellent manager is the one who can predict, prevent and resolve issues before they arise, only involving the senior leadership when absolutely necessary. In this metaphor, our conscious awareness is 'senior leadership' and allostasis is the 'excellent' manager.

A system designed to function based on predictions makes a lot of sense, although for centuries the dominant theory of neural activity was that it primarily functioned based on response to stimulus. This presents the human brain as dormant in the absence of stimulus, like Siri waiting for someone to say something that might sound a little bit like 'Siri'* and snapping to attention. But if our body functioned in

* Such as 'necessary', 'serious' and 'avocado'.

strict stimulus-response fashion, we would be living in a state just shy of death. We'd have to slip into dehydration to remember to drink, or flop to the brink of starvation to eat. A system built on responding to needs, rather than anticipating them, would be disastrous: life would be a series of surprises, constantly responding to crisis – exhausting and incredibly inefficient.

Efficiency, then, requires learning to predict the needs of the body in order to anticipate and meet them in the future. How do we do this? By constructing and running our own internal model, or simulation, of our lived reality, informed by our past experiences. In order to do most things – catch a ball, throw a punch, put one foot in front of the other – some part of us needs to know and prepare for what's going to happen next. Over time, we get so good at predicting what's about to happen next that we don't need to pay attention to what's actually happening 'out there', meaning our predicted model – and the sensations it generates – becomes our perceived reality. Our sense of the world is often generated right in our own minds – it's the Matrix. But this Matrix is not designed by our robot overlords, it's actually designed by us.

According to a whole bunch of highly regarded neuroscientists, this is how our brain is organised to function. Lisa Feldman Barrett and fellow researcher Benjamin Hutchinson wrote in a 2019 article, 'Unlike psychology's traditional framework, in which perception and action are separate processes with one causing the other, the predictive-processing approach suggests that perception and action are united by the brain's internal model in its effort to efficiently navigate its body in the world.' This internal model of our world is always just milliseconds ahead of our conscious awareness, allowing us to predict and prepare.

The synchronicity between our internal model and what happens in the 'real' world is key in constructing our sense of agency. When our predictions either don't match what we do (we gave the command to move our arm, our arm did not respond) or what we predicted would happen doesn't (our fist landed on our opponent's neck instead of his face, or the predicted pain is far greater due to an opponent's lead jaw) we don't feel in control. When our predictions are wrong we have three options to try to regain a sense of control and resolve the dissonance: we could adapt in the moment and update our model to accurately reflect the afferent signals coming in; we could try to change the circumstances or environment to match our prediction; or we could filter out the conflicting signals and carry on.

Prediction through simulation, also called perceptual inference, carries all kinds of implications for how we experience pain (and our entire lived reality). An example of our predictive processing in action is illustrated in what is called perceptual sensory attenuation, and it's a big reason why the pain we expect is not as bad as the pain we don't (and why we can't tickle ourselves). Consider this scenario: milliseconds before Ibrahim throws a right hook, his internal model had already played it through and generated the necessary series of neurochemical signalling required to not only swing his arm but also to brace in anticipation of landing the blow. This internal model of outflowing signals is called our 'efference copy', and research shows it essentially can work to cancel out, or filter, the resulting reafferent signals or the expected afferent signals – stopping them before they reach the level of our consciousness. Basically, it's as though the nociceptive signal that arrives at the dorsal root ganglion from our knuckle after it connects with our opponent's face is met with our very efficient manager saying, 'Yep, already got it, in fact – here's some endogenous opioids, no need to trouble leadership.' This filtering helps promote perceptual stability and seamless movement, and it means that our nervous system is proactive in responding to anticipated needs.

Simulation is essentially neuronal firing based on best guesses of what is about to happen next – from where you're moving to what you are hearing, seeing and feeling. It's powerful; studies show, for example, patients expecting morphine generate their own endogenous opioids in advance of delivery of the real thing. And if our mental model can generate powerful placebos, then it can – and does – simulate nociceptive signals, making us feel pain that, in a sense, isn't 'really' there. This is the nocebo effect and it has been widely observed, not just in dramatic cases like the builder who stepped on the nail, avoided the puncture but didn't dodge the pain. For example, women who expected severe post-operative pain following surgery related to breast cancer reported more pain in a 2017 study compared to those who didn't expect the pain to be serious.* Because of our efficient simulation model, when we expect to feel pain we generally will. The placebo and nocebo effects are terrific examples of predictive processing in action (how we can use these effects is something we'll get into in a bit).

Another tremendous illustration of how much of our sensory experiences are generated in our mind – no body needed – is in the rubber-hand illusion, easily the most fun you can have in a lab with a fake arm. In this experiment, through a phenomenon called proprioceptive drift, people come to feel ownership of a prosthetic hand after a series of visual and tactile stimulations. The volunteer is invited to place their right hand on the table in front of them next to a realistic prosthetic left hand; their left hand under the table, where the researcher can still reach it but out of sight of the volunteer. After asking the participant to keep their eyes on the fake hand, the researcher begins to brush both the fake hand and their real hand, which is under the table, simultaneously. After a few visual confirmations, our internal model is updated: the fake hand now belongs to us. And we're not just adopting the fake hand as a kind of third appendage – we seem to also abandon our real one, to the point that motor signals to the real hand actually decrease.

The rubber-hand illusion also underscores the point that in order to ensure accurate predictions, our internal models have to be as close to reality as possible. If Ibrahim's internal model didn't include a prediction of nociceptive pain and potential tissue damage, based on his past experiences, he'd be completely unprepared and he could make bad decisions, such as hitting far harder than his bones could withstand.

Perhaps one of the more interesting – not to mention potentially therapeutically useful – illustrations of the power of our predictive

* Likewise, telling a child not to worry before a painful experience results in just the opposite. One study from 2010 found that children exhibited more fear of painful procedures when parents tried to be reassuring than when parents simply tried to distract the child. This, researchers agreed, was counter-intuitive, but repeated messages of assurances not to worry only increase the salience of worry for the child. They hear 'don't worry' and, thanks to our predictive brain, the child prepares by constructing an instance of worry.

internal model is in virtual-reality (VR) therapy. Up to 90 per cent of those who undergo amputations report sensations, often painful or irritating, in a limb that is no longer there; most of those people have found that drug therapies, nerve blockers, surgery and physical therapy do not provide relief. In the 1990s, neuroscientist V.S. Ramachandran of the University of California, San Diego began working with individuals who had lost a limb after a period of paralysation. He theorised that there could be a way to 'trick' the brains of patients suffering from phantom limb pain into seeing that limb as restored and moving, loosening up the discomfort.

Ramachandran took a cardboard box, removed the lid, and slid a mirror vertically down the middle. He cut two holes in the front of the box, one on either side of the mirror. The first subject was asked to place both arms – the whole limb and the 'phantom' limb – through the holes, and to align the whole hand with where he felt the phantom one to be. Ramachandran then asked him to move both his arms in concert, as if he were conducting a symphony, and watch his reflected 'left arm' move. It worked; with his internal model reset, the pain diminished and eventually the impression of the phantom arm disappeared.

With advances in virtual and augmented reality, our ability to manipulate a sense of body ownership is more advanced than ever.* A study from 2018 published in *Frontiers in Neurology* tested a game-like VR training with two study participants who had previously undergone amputation just below the knee. The participants were represented with a VR avatar controlled by sensors placed on their hips and joints; when the user moved so did the avatar. Participants played 20 minutes of a VR game, which required significant movement of the amputated limb, followed by a free choice of games for 40 minutes. According to the researchers, both participants reported a significant decrease in pain ratings, while their baseline pain ratings also went down over the course of sessions.

The more sensory integration the better. An interdisciplinary group of researchers from the University of Maryland conducted a

---

* The global augmented and virtual reality market is poised to hit nearly $19 billion in 2020 and is expected to continue to grow; fake reality is only going to become more real the more money we throw at it.

comprehensive review of the effectiveness of VR and music for pain management. Their findings, published in the September 2019 issue of the journal *Pain*, point to VR reducing the intensity of acute pain, and that when combined with multimodal stimuli like sound, body haptics and immersive environments, the results are even better.

One of the biggest hopes for VR is that it will improve physical therapy outcomes for motion-related pain by disrupting the predictive model that reinforces guarding behaviours. Guarding behaviours are formed through basic classic conditioning. Say you were rear-ended in a car accident and left with a bad case of whiplash. Every time you turned your head, pain struck your neck like a lightning bolt. After only a few attempts, you'd quickly learn not to turn your head, not to even try to move it. Soon, any kinaesthetic, or movement-related, cues – the slightest tilt of your chin as you prepare to look over your shoulder, nodding yes, shaking your head no – automatically raise the red flag to stop. Over time, as the relationship is reinforced, patients restrict their range of motion more and more, but it's not the pain itself that sustains the avoidance, it's the fear of it. Many patients may not even know how their posture and range of motion has adapted until the consequences of those adaptations start producing their own pains. While guarding behaviours might protect patients from pain in the short term, they lead to muscle weakness and more pain in the long term.

Essentially, the idea of VR therapy for motion-related pain is to bypass reflexive, fear-based guarding movements by tricking the brain. In a VR environment, patients would adopt an avatar body as their own, and through adjusting the VR software, the user's perception of their range of motion would be distorted. For example, the user may think they are only lifting their arm 45 degrees because that is how high their avatar arm is raised, but in reality, they may have lifted it 70 degrees. With practice, the patient would begin to move further, regaining a fuller range of motion.

So far, the evidence supporting this approach has been mixed, but it also highlights that reduced pain is not always the best measure of progress. A 2015 study reported in the journal *Psychological Science* and led by Daniel Harvie from Griffith University in Australia confirmed that manipulating patients' perception of how far they were rotating their head increased their range of pain-free movement. This was an encouraging finding; Harvie followed up with an experiment comparing pain ratings during head rotation among those with persistent neck pain in a perceptually accurate VR environment, and an 'enhanced' VR environment. Though there were no significant differences in pain ratings between the environments, there were other therapeutic benefits: seven out of the eight participants scored lower on the Neck Disability Index after the interventions.

Finding control by looking inside

When fear of pain intensifies the experience of pain, it reinforces that pain as terrifying. Avoiding that pain becomes a top priority; individuals restrict any movement or activity associated with the pain and become hypervigilant to any physical cues that might indicate the pain is coming again. Thinking about pain becomes pain catastrophising – the act of describing or thinking about pain in amplified terms, ruminating on it – and it has negative effects on how we perceive the pain we're in, as well as the pain we might be in. This is true for acute pain, it's true for pain during the healing process, and it's true for pain that accompanies chronic conditions such as heart disease.

Heart disease is the number-one killer in the US and the UK. There are a number of different forms it can take but the most common is cardiovascular disease, the narrowing of blood vessels that slows or stops blood flow to the heart. Its symptoms are terrifying: angina pectoris, as it's called, is the painful constriction of the chest accompanied by a struggle to breathe and the certain conviction that you are dying. It is a pain cave of the most mortally terrifying kind – dark, inescapable and utterly outside of your control. But what makes angina so confounding is that you might not be dying after all; the severity of the pain is no indication of the severity of the heart disease. Your angina pain can be unimaginably intense, even when there is only a mild blockade of the arteries feeding the heart. This pain can be chronic, recurring with stunning regularity, and every time it comes with the knowledge that this time could be it, this one could kill you.

Dr Austin Leach is a cardiac pain management specialist near York, England, and he sees this pain a lot. 'When people get this crushing feeling in their chest… they become frightened. And if you're frightened, it makes things seem more exaggerated. And being frightened and in pain, it's worse than being just in pain,' said Leach, also a spokesman for the British Pain Society, a group of medical professionals brought together by their commitment to understanding and managing pain.

Out of the thousands of signals that make it to our CNS, the ones that tax our body the most are going to be prioritised; pain and fear are at the top of the list. Either on its own can drive our sympathetic nervous system into high gear; when they are combined the response is amplified. This is particularly concerning for heart patients: an increase in respiration can tip into hyperventilation or otherwise restricted breathing, further slowing the flow of oxygen-rich blood to the heart muscle. The pain intensifies and, panic mounting, these patients do what their doctors told them: go to the hospital. 'When they get to hospital, they feel safe, so [the pain] calms down. And the A&E nurse says "Oh, it's you again, we've kept your trolley warm." It doesn't make people feel any better, it makes them feel like a nuisance,' said Leach. After a while, this response becomes almost a reflex, a neural pathway dredged by experience, and the cycle of pain, panic and social chastising degrades their quality of life.

While avoiding pain is reasonable, assessing a patient's fear of pain offers real insight into how they will likely manage their pain. One of the most reliable measures of fear of pain is the Pain Catastrophizing Scale (PCS), developed in 1995 by Michael Sullivan at the McGill University Centre for Research on Pain and Disability. The survey consists of 13 questions that start with 'When I'm in pain…' and then a series of possible thoughts and feelings that patients can rate from 'not at all' to 'all the time'. These include thoughts such as how much they worry that the pain will never end, think about other painful events, and fear pain will get worse. Numerous studies show PCS scores are a reliable predictor of pain outcomes: a study of patients with a chronic pain condition found that high PCS scores were correlated to patients experiencing more distress pre- and post-surgery, and more intense pain days later.

When an angina patient fears an attack is imminent, especially with the knowledge that coronary disease is often fatal, catastrophising doesn't seem like a choice but rather an instinctively appropriate response. This fear has kept some of Leach's patients imprisoned in their homes, terrified to leave in case of an attack. It's his job to open the door for them. 'We sit patients down and say, "You've got yourself into a bit of a hole, we're going to explain how you got there,"' he said, explaining that they work with patients to help them understand what catastrophising is and how it hurts them. Some of what Leach does is in line with how endurance athletes cope with pain in sport, helping his patients recast the painful episode as non-life-threatening. Leach gets them to see the light at the end of what seems like a constricting tunnel 'using high conscious behaviours or logic to push to one side the unpleasantness of the now.' This is coupled with what boils down to meditation and breathing exercises to help sufferers relax. 'By using simple physical repetition, non-challenging movements,' said Leach, 'you can lull the mind into a state of not caring, if you like.'

There is a growing body of research underscoring the benefits of similar meditation and mindfulness techniques in pain management; these practices have roots that go back centuries. These practices persist because learning how to become more aware of oneself and grounded in the moment can be a powerful tool to manage stress, pain, and whatever else life throws at you.

Right now, your body is brimming with activity. Much of it you're unaware of, such as digesting your dinner, rebuilding muscle from yesterday's workout. Some of it you're barely aware of, such as the sound of traffic in the street, and some of it you're very aware of, such as the fact that you are now reading this sentence. The process of making sense of all these signals – or, more specifically, our brain's representation of all this noise – is called interoception.* Though much of this activity is beyond our ability to perceive, we can 'bring awareness' to a lot of what's happening in our body through what's called interoceptive awareness. How 'tuned in' people are to their own physiological state is referred to as interoceptive sensibility and can be measured with simple questions like 'How often do you notice when your mouth is dry?' or 'Do you often notice when your muscles are sore?' This trait is variable – some people are not very connected or concerned with what's

* It works in conjunction with allostasis, which regulates and predicts based on interceptive signals.

happening in their body, while others may have a hard time focusing on anything *but* what is or could be going on in their body. Being aware, however, doesn't also mean being right, and the degree to which we're more or less accurate about what we think is going on inside our body is referred to as interoceptive accuracy.

Finally, our highly evolved human brain allows us to do more than just perceive, with more or less accuracy, what is happening in our body; we are gifted, or cursed, with the ability to think about, reflect on and judge our own interoceptive abilities. This metacognition is interoceptive awareness and reflects the extent to which we are confident that what we think we're experiencing is actually happening. Typically, these three traits are significantly correlated: someone who is highly sensible of their interoceptive situation is often more aware of their physiological states and is usually more accurate about those states. Conscious interoceptive sensibility, accuracy and awareness exist along a wide continuum that varies between people and can change from one moment to the next. But they are traits that we – like Leach's patients, like the athletes we met in this chapter – can practise and improve.

Developing interoceptive awareness can powerfully impact how we experience pain in the moment, which in turn influences the choices we make. Here's an example: you ask your friend Liam how he's doing. He replies, 'Meh.' Liam doesn't elaborate, nor does he seem to reflect beyond that; he has low interoceptive awareness. You ask your friend Michelle how she's feeling. She replies, 'A bit dreadful, actually. My head aches and my mouth is dry – I think I'm dehydrated – and my muscles are a bit sore. I'm also a little sad today. Maybe I'm coming down with something?' Michelle's self-inventory, demonstrating a high level of interoceptive sensibility, reveals that she might be getting a cold. She then decides to make a point of drinking some water, taking some ibuprofen to manage any inflammation and possibly pre-loading some vitamin C.* But it wasn't *just* that she drank some water and took

* Does vitamin C actually keep a cold at bay? Nope. According to Cochrane reviews, there is no evidence to support the claim that vitamin C shortens the duration of colds, nor prevents them from coming on. And yet, we persist – Linda's husband is a big believer in vitamin C for colds, despite living with a terrible sceptic with access to medical journals. Behold the power of faith.

some pills. Michelle also demonstrated high emotional granularity, being able to identify and name her sensations with specificity. This is an important part of interoceptive awareness and improves emotional regulation, including pain modulation. In fact, researchers are increasingly pointing towards dysregulation of interoceptive processing as a shared, likely key component in many disorders, including depression, PTSD, sleep disorders, autism and addiction. When you know what your body is saying, and have an idea of why it's saying it, you have a choice about what to do next – you're in control. Developing these skills extends beyond the self; evidence shows interoceptive awareness increases compassion and pro-social behaviour.

One way to develop and practise the skills of interoceptive awareness is with mindfulness meditation (MM). MM is practising interoceptive awareness with intent, although how that is accomplished varies significantly among teachers, spiritual guides, clinicians and researchers. There's Open-Monitoring (OM) meditation, which teaches practitioners to focus attention on internal and external sensations in an accepting and non-judgemental manner. This might look like feeling the ache in your ankle without following up with thoughts like 'I shouldn't have run that extra mile' or 'My stupid bones are melting away'. The Body Scan is another popular technique, in which practitioners sequentially inventory each part of their body and notice the sensations, often practising 'letting go', attempting to relax the muscles.

Measuring 'bringing awareness to' and 'letting go' in a human is a challenge although not impossible with the help of new technologies including fMRI, older ones such as EEG, and by measuring heart rate, skin conductance, and muscle contractions. It's getting funding for this type of work that's the real hard part, especially when competition for research funds prioritises interventions with strong commercial potential (ahem, drugs). But peer-reviewed research into the power of these different techniques is growing, and the findings relevant to improving our relationship with pain are promising. For example, Focused Attention meditation, which consists of practising rhythmic breathing similar to what Leach teaches his patients, has proven helpful in decreasing pain's unpleasantness and improving stress management. A 2014 study found that Marines who received 'mindfulness-based mind fitness training', emphasising interoceptive awareness, exhibited significantly better stress recovery following the combat session than Marines who hadn't had the training. And, in a measure of pain tolerance, a 2014 study found that those who adopted yoga-based practices were able to tolerate pain twice as long as matched controls.*

Leach is using mindfulness techniques to give patients a sense of agency in an experience that previously felt entirely outside their control, helping patients treat their pain by addressing and reducing their fear. In a 2013 article in the *British Pain Society Journal*, he and a co-author explained that the therapy seems to help the cerebral cortex learn 'how to make sense of the clamour of noise' coming from a busy sympathetic nervous system 'that is doing its best to control a highly complex interactive system.' Leach suggested this keeps the systems responsible for pain modulation from 'jumping to the wrong conclusion' about how to treat this particular episode. 'Not only are there fewer emergency hospital admissions, once patients have been educated about angina and ischaemic heart disease and they become more confident in dealing with painful episodes, but they also experience less frequent and less severe pain, take lower doses of fewer drugs and show a reduction in the rate of myocardial infarction.'

Anecdotally, said Leach, the patients who respond to the treatment say they feel like they've got their lives back. He recalled a follow-up meeting with one man who'd barely left his home in years. 'We asked

* We have to note that bringing attention to physical sensations sometimes makes things worse. Neuroimaging studies have shown that individuals struggling with the impact of trauma have more active interoceptive networks compared to non-traumatised populations. Our interoceptive network is critical in the process of comparing current information to previous experiences in order to anticipate needs and formulate predictions. In instances where the sensations tap into previous trauma, the elicited feelings may be overwhelming, resulting in dissociation, fear, loss of control or panic. This can leave some traumatised individuals feeling nervous to fully engage with the sensations of their body, that it is not 'safe'. This underscores the importance of teaching supportive techniques and not throwing survivors into the deep end because, with support, practising interoceptive awareness can help trauma survivors re-engage with their body and once again experience it as not only safe but fully integrated, which in turn is related to greater well-being.

him how he was doing. He said, "Fantastic! I can't explain it, Doc, but this is how it is, it was like I was in prison,"' Leach recalled. The man described his pain as the warden, but said that now, he bangs on the bars of the cell to bring the warden around, just to remind him who's the boss. '"I've got this breathing and I bring my angina on, so I can make it go away!"'

The promise – and limitations – of control

Athletes like Adharanand Finn challenge their bodies to endure pain and confront it as a malleable and impermanent concept. Finn can stop running at any time, but he doesn't because pain is something he can transform altogether and on his terms. Rami Ibrahim can't control his opponent, but through preparing for the pain, *expecting* the pain and confronting his worst fears he retains control of himself. Austin Leach's patients find their way out of a desperately terrifying experience that feels – and often is – out of control, by focusing on what they can control, even if that's only their own breath.

Preserved agency in the face of threat is the difference between getting stuck in the pain cave and emerging into the light. Of course, there are limitations. We don't always have the choice to stop running or to tap out, and we cannot always 'reframe the pain away' or count on our expectations and endogenous opioids to carry us through. Some pain will be horrifying and debilitating. But we can learn – from Finn and Ibrahim and the man who brings on his angina pain just to make it go away – to find or create the spaces where we can assert our agency, even if it's in proclaiming to the world, 'THIS IS MY BODY! THIS IS MY PAIN!' In those moments we take away the power pain draws from our fear, the parts of pain that feel bigger than we are. This doesn't mean giving up or giving over to suffering, it doesn't mean abandoning hope. It means owning pain: you're a body in pain right now but you're also a body *not* in pain in the future.

Choosing to push our bodies, whether it's on the track or on the mats, in the water or the ring, gives us an opportunity to practise pain, to *feel* our bodies, sense the changes and decide what meaning we want to make of them. And sometimes when it seems like our bodies are shouting 'Stop! This hurts!' maybe we need to listen closer to hear the part that's whispering, 'Oh, go on then. You can do a bit more.' It's not easy; change is daunting, and feeling our bodies can be scary. It's also pretty scary to learn how much of our reality is mediated by our predictive processing, our running inner model formed from previous experience. But in recognising this we find opportunity, through practising interoceptive awareness, reframing and recategorising to update our predictions, to rewrite the code of automatic processes that aren't doing us any favours, that are *causing* pain. With time, we can turn practice into prediction and, in so doing, take back control.

CHAPTER SIX

The 'Shocking' Truth: Pain is Useful

One block from the bustling Union Square market in downtown Manhattan, a middle-aged white man sat at a computer in a windowless office on the sixth floor of a high-rise. Thinning sandy-brown hair, slim with a slight paunch, wearing a short-sleeved blue and green striped shirt and beige chinos, he looked like a stock photo of an 'office worker'. But this guy – we'll call him Stan – wasn't here to fix the internet connection. He was here to participate in a study being run by Ashley Doukas, then completing her PhD in the New School's clinical psychology programme. For her dissertation, Doukas was looking at whether and how people used pain to regulate emotional states. And Stan was about to shock himself with enough electricity to take down a charging Rottweiler. On purpose.

Stan sat facing a computer monitor, keyboard and mouse on a black desk, the only furniture in the room. Several wires disappeared under his shirt, connected to two electrodes placed just below his bottom left rib, measuring his heart rate. On his head sat what looked like the plastic frame inside a hard hat – this was the mobile electroencephalogram, or EEG, which measured his brainwave activity through clusters of electrodes pressed against his temple and scalp. His left hand was hooked up to sensors that monitored his sympathetic nervous system activation, the system that kicks in when our body becomes stressed or aroused; this one in particular was monitoring Stan's galvanic skin response, or in plain English, sweat. Whether or not Stan was sweating, how fast his heart was beating, what his brain activity looked like, would all give Doukas a clearer picture of the kinds of changes that were taking place in Stan's body while he decided whether or not to shock himself. Stuck to the skin of his inner left forearm were two square white pads that would deliver the electric shocks.

Stan spent about 20 minutes wired up like this, clicking through a series of tasks and psychological surveys screening for depression and anxiety, along with a battery of questions about his personal history. Once he'd completed this section, a woman's voice spoke through tinny computer speakers: 'We're asking you to sit quietly in the room until we come and get you. You will not be able to look at your phone, but you are free to think about anything you want. Or you can choose to administer either a high- or low-level shock for one to five seconds.' The screen in front of him went white, then two rows of five black boxes appeared. The top row was labelled 'high', the bottom 'low', and each box was numbered one to five. 'High' or 'low' indicated the level of electric shock, the numbers indicated the duration of the shock; box one was one second long, box five was five seconds long. 'Do you have any questions?'

'No,' Stan responded. And immediately he began clicking the boxes. One second, low. Three seconds, low. Five seconds, low. One second, high, which made him jump a little. Three seconds, low. Three seconds, low. Three seconds, low. One second, high. He found a groove: repeated low-level shocks for three seconds, with the occasional high-level shock for one second thrown in.

It's not as if Stan didn't know the shocks were, well, shocking. The strength of the shocks was calibrated for each participant at the beginning of a session; subjects were able to test and decide for themselves what high and low meant. In this case, the lower end was in the same arena as an intense electro-static shock – think pulling off a jumper in winter – tolerable but not pleasant. The high was like bumping into an industrial electric fence or grabbing a three-pronged plug at the base (in America, at least, where the voltage is lower). These shocks were low frequency, associated with nociceptive reactivity – in other words, they hurt. And throughout the 10-minute session, Stan clicked steadfastly on with few pauses, jumping only rarely, even as the muscles in his forearms visibly contracted.

Three doors down, Doukas and three of her undergraduate assistants watched a wall of monitors. Two monitors showed footage from the closed-circuit cameras trained on Stan; one showed a screen mirroring Stan's own screen; three others displayed his heart rate, his GSR and his brainwave activity respectively. All eyes moved between the mirror of Stan's monitor and the three showing in real time how his brain and body responded to his choices.

Doukas's hypothesis was that if pain is a fast and effective method of sympathetic nervous system activation, then under the right circumstances and when it's under our control, pain can be adaptive, helping us get what we need. In order to find out whether and how people can use pain in an adaptive way, we need to know what's happening in the body and brain when its owner chooses to induce pain. But getting approval for the protocol was a challenge. She designed the study around a 'normal' population, meaning there was no diagnostic-specific criteria required for recruitment or enrolment. The Institutional Review Board (IRB), which oversees all research, didn't seem to get this. No one from the general population would voluntarily inflict pain on themselves, they said, and if they did, they were not asymptomatic and therefore should be excluded from a study on a 'normal' population. The IRB's response isn't surprising: culturally, the prevailing assumption is that self-induced pain must be a symptom of a mental pathology.

Doukas fought back. 'Many people, not just clinical populations, use pain to change their emotional states or even experience pleasure. There is nothing inherently unhealthy or even unusual about using pain this way, we just rarely recognise it since it seems normal to us or we don't think of pain as a primary motivator for engaging in the behaviour,' she told us. 'For example, we might think that someone who likes to work out tolerates pain to get results, but maybe some people actually just like to feel the burn, *and* also enjoy the result.' Not to mention, most people have at least once pinched themselves or slapped their face to try to stay awake, picked at their cuticles when they're stressed or bit their lip to keep from laughing at inappropriate times. The IRB relented after she reminded them how humans have been using pain in healthy, 'normal' ways for centuries, and that it is a norm in cultures around the world.

Back in the research room, Stan was still administering shocks to himself. And he certainly wasn't the only subject who had. 'So far, every single participant has chosen to shock themselves, most multiple times,' Doukas told us that day. At the end of his 10-minute session, Stan sat dazed. He watched mutely as one of the research assistants removed the gear from his head, and gently pulled the sensors from his forearm, fingers and chest. As she detached the multiple wires, she said, 'So, we saw that you pretty consistently chose the low-stimulation shock. What made you choose the one- to three-second pulse instead of five?' He didn't respond for a moment, maybe thinking or still dazed. 'I don't know,' he replied.

Stan might not know, but thanks to research like this, we're all getting a little bit closer to the answer. The findings from Doukas's study were published in 2019 in the American Psychological Association's peer-review journal *Emotion* under the title, 'Hurts so Good: Pain as an Emotion Regulation Strategy'. There were actually two parts to Doukas's study, both involving electric shocks; Stan had already completed the first task, which involved viewing disturbing images such as car explosions, gun violence and abject poverty. The explicit goal of these photos is to make people feel bad when they see them; participants were shown an image and then offered a choice of coping strategies they could use to make themselves feel better or at least regulate their emotions. These strategies were avoidance ('think of something other than the picture'), reframing ('change the meaning of the picture in your mind so that you feel less negative about it'), self-administering a high-stimulation electric shock or self-administering a low one.

The results were astonishing. Among participants, 67.5 per cent chose the painful stimulation as a coping mechanism to deal with a negative image. And in the second task, when participants could choose to shock themselves any time they wanted, most did – in the final assessment, 71.2 per cent of subjects self-administered low-stimulation shocks at least once, and 59.3 per cent self-administered high-stimulation at least once. One person even gave themselves a high-stimulation shock lasting 345 seconds, more than half the duration of the entire task. There were no significant differences by gender: 65.8 per cent of women and 57.1 per cent of men chose high stimulation at least once across the two tasks. And there were no demographic predictors of behaviour; the participant population was representative of the general public, with an average age of 35 and representing diversity of race, income, marital status and education.

Doukas also measured whether choosing the high stimulation was related to how people felt before they even started, the idea being that pain is a last-ditch effort for people who feel just awful. But the data showed no relationship between how people felt before the trials started and whether they chose to shock themselves, demonstrating that intense negative emotion is not a precondition to using nociceptive activation to feel better. This point alone shows the importance of conducting these types of experiments on 'normal' populations.

The results of her research challenge many of the clinical, social and medical assumptions about self-administered pain. Primarily, it dispels the notion that self-administered pain is pathology, demonstrating that healthy populations can intentionally use pain to cope. As she told us, 'Everyone is doing it, and it's not just about avoidance or boredom.' That's a nod to the prevailing explanation for this behaviour, made popular by a 2014 study similar to hers. In a paper titled 'Just think: Challenges of a disengaged mind' psychology researchers from the University of Virginia and Harvard argued that boredom was the reason so many of their participants chose to shock themselves during a task that involved sitting in a room alone with no external stimulation. That paper earned a lot of mainstream media attention, largely because of the gratifying, *schadenfreude*-inducing narrative that millennials hate being alone with their thoughts so much that they choose pain over contemplation. More recently, a 2016 study from researchers at the Maastricht University in Belgium seemed to confirm this: they found that subjects were more inclined to give themselves electric shocks when they were bored than when they were sad.

Doukas acknowledged that there may be a link to boredom but she offered a different perspective. After all, what is boredom but an unpleasant, low-arousal state? On the affective circumplex we described in Chapter 1, it would be solidly left of centre, hovering in the negative area. Thinking of emotions as goal-related, feeling bored generally precedes an attempt to *not* be bored.* One way to do that, especially if you're a participant in a study that has given you the tools to do so, is to shock yourself: '[If] boredom is conceptualized as an unpleasant low-arousal state,' she says, 'we may interpret the choice to shock as evidence of up-regulation.' The question of whether today's

* Unless you're an eight-year-old boy, at which point a claim of being bored usually precedes another, louder claim of being *boooooored*.

generations experience more boredom than in the past is an entirely different one, and not the point: if we don't like boredom and want to do something about it, pain is *something*.

Doukas's data also confirm that self-administered, non-injurious painful stimulation is an effective emotional regulation strategy. As demonstrated in the first task, negative emotions decreased after self-administering high stimulation – pain didn't make people feel worse, it made them feel better. In fact, it was as effective as avoidance, which is ironic considering the prevailing attitude towards pain *is* avoidance. And, the big surprise: pain proved more effective in increasing positive emotions and decreasing negative emotions than cognitive reframing, long considered the gold standard of emotional regulation strategies.

But why are we surprised? As Doukas noted, intense sensation – pain – has been used therapeutically by cultures around the globe for centuries. And in the previous chapter we talked about how people accommodate pain in endurance sports. However, we didn't really explore the motivation behind that – in fact, the promise of pain might be part of why we do it. Ultra-runner Adharanand Finn told us, 'I definitely had people telling me that they relish the pain, that that was why they ran, that they were looking forward to that confrontation, that crisis.' Though pain wasn't the primary draw for him, he said, 'There was something oddly satisfying about being in the moment of crisis, being in pain, that I was surprised by.' Meanwhile, runners on Instagram love to quote psychologist Jacqueline Simon Gunn: 'The pain of running relieves the pain of living.' This makes sense, especially if some of the pain of living results from anxiety. Running, along with most activities that kick arousal into high gear, forces neural networks to reprioritise activity, pulling resources away from higher-order executive functioning. That experience is as much a product of the healthful benefits of running as the pain it induces.

The fact that it is surprising to many – not least the IRB – illustrates just how siloed the word 'pain' has become. Pain is supposed to be bad, always something to avoid, but this negative paradigm blinds us to its potential therapeutic uses and treats people who might find relief using it as pathological. Pain *can* be useful. However – and it is a very big however – this is not to downplay the seriousness of what's called 'non-suicidal self-injury', or NSSI, the practice of injuring oneself without the intent to end one's life. We differentiated between 'good' pain and 'bad' pain in the context of childhood development in Chapter 4; similarly, the line between pain that can be useful but not harmful and NSSI is an important one, but one that scientific research is only recently trying to draw.

NSSI is a very real public health concern, and it is dangerous. Young adults between 18 and 25 are considered the most 'at risk', but overall prevalence is estimated at somewhere between 4 and 28 per cent of the population. The most frequent form of self-injury, accounting for 70 to 90 per cent of individuals, is skin cutting, but other less common practices include banging or hitting oneself, scratching, burning, or interfering with wound healing. The physical dangers associated with clinical NSSI are not trivial. The risks vary in accordance to the seriousness and frequency, especially among those (usually adolescents) who may not be aware of the acute consequences – the risk of cutting too deeply, for example – or long-term effects such as scarring, tissue damage and an increased risk of infectious diseases. But one of the biggest issues in talking both about NSSI specifically and in talking about pain as emotional regulation is the lack of clarity around exactly what constitutes pathological NSSI. This also explains why estimates of prevalence of the behaviour in the general population vary so widely. The latest edition of the *Diagnostic and Statistical Manual of Mental Disorders* (*DSM-5*), describes six criteria for a diagnosis of 'non-suicidal self-injury disorder' (NSSI-D). The criteria, however, fares poorly in real world validity testing; for one thing, the entry doesn't offer a list of what types of behaviors are considered NSSI, only notes that they can't be 'socially sanctioned'.

But just because a behaviour is not socially sanctioned does not in itself make it problematic – it wasn't too long ago that masturbation was socially disallowed, self-injurious behaviour meant to cause 'insanity'.* Meanwhile, the criteria that the behaviour must cause 'clinical interference/distress' – the one typically used to suss out bad habit from pathology – is problematic: those who self-injure report the practice as helpful, though certainly distressing to their friends and family. Perhaps unsurprisingly, even the *DSM-5* agrees the NSSI-D is a 'condition requiring further research'. Especially more research approaching the behaviour from a mechanistic rather than moralising perspective; much of the research in the 20th century treated NSSI as failed attempts at suicide or efforts to manipulate others, rather than what it is – a specific kind of emotional regulation behaviour.

NSSI is one small, albeit concerning, part of the landscape where physical sensation is used for emotional regulation. And this is why the fact that Doukas recruited among a 'normal' population is so important; so much of what we know about how pain functions to regulate emotions stems from comparing those who use NSSI (variably defined) to those who don't. There are important, significant differences that we will get into later, but in narrowing the conversation we miss opportunities to learn how people are using pain in everyday, common, non-injurious ways and, more significantly, we miss investigating pathways that might lead to healing for those who do turn to NSSI.

Meeting pain with pain

Karl Marx may have been the one to say, 'There is only one antidote to mental suffering and that is physical pain', but it's Aiden Carter† who lives it. As an adolescent, whenever he was sad or angry, Carter cut himself. He didn't, he says now, want to harm himself. 'I was able to cut myself and feel something,' he explained. 'After I cut myself I could visualize that as the wound healed, the emotion that caused me to do it was also healing. It became a physical and visual representation of what was going on emotionally for me.' But when the people

---

* To be frank, the devices sold to the public in the 19th and 20th centuries to curb this terrible habit seem more likely to cause actual harm. Google 'spermatorrhoea ring' for more grisly details.

† Names have been changed to protect the individual's privacy.

around him started telling him that this was self-destructive, worrying behaviour, he stopped; depression followed.

Carter soon began using drugs, suffering a heroin overdose at 19. 'I recognized that I needed to change something in my life,' he said. 'Not wanting to half-ass anything, I ended up serving.' Carter, who was born in 1970 in Canada, joined the Canadian Army in his early twenties and soon after was deployed to Iraq during the Gulf War. He served for five years full-time and then part-time for another six. Carter kept away from drugs, but the spectre of depression was always hovering. Following his time in the service, he again found relief by turning to his body. Only, this time, Carter sought a safer, more socially sanctioned avenue – body piercing and modification. And the deeper Carter went into his practice, studying to become a body modification artist, the more he kept coming across articles about and videos of a man who called himself Fakir Musafar.

Musafar was born Roland Edmund Loomis in Aberdeen, a small city in South Dakota, in 1930. His father was a mechanic, his mother was a homemaker and both were Lutherans who hoped their boy might become a minister one day. According to the *New York Times*, young Roland had other plans: from an early age, he was obsessed with learning about the rituals of 'primitive cultures' through the pages of *National Geographic* magazine. At 14, he secretly pierced his own genitals in the basement. Loomis's path to becoming Fakir Musafar was a secretive one – he practised corseting, self-bondage, piercing and tattooing while getting a degree in education, serving in the army during the Korean War, earning a masters in creative writing, teaching ballroom dance and becoming an ad exec. In 1977, however, Loomis came out as Fakir Musafar at the first International Tattoo Convention in Reno, Nevada, finally sharing with the world the spiritual joy he'd found through body modification, bondage and branding – what he came to call 'body play'. Musafar's new life made him the founder of a movement, one that promoted ecstasy and emotional release through actions that most people would describe as painful. 'There is no pain,' he told *The Vancouver Province* in 2005. 'There is only intense sensation.'

Musafar's philosophy resonated with Carter, who decided to take one of Musafar's courses at his training centre in San Francisco.

Musafar became Carter's mentor, and one of the most important influences in his life. Over the past decade, Carter has lead and helped to organize hundreds of gatherings, with Musafar's philosophy as a guiding post. When they first met, Carter said Musafar gave language to the feelings and truths he felt but could never articulate. 'Pain,' Carter explained, 'is getting up in the middle of the night and kicking the corner of the bed on the way to the bathroom … [I]f the sensation is something that you expect and embrace, it's not pain.'

This is not to say that some of his body stress practices, which include flesh hook pulls and suspension, don't hurt. 'The piercing hurts every time, there's no way around that,' says Carter. But it's an integral part of the emotional process, one that ends in healing, joy and release: 'To me, that's like the bus. I need to get on the bus to get where I'm going. That is the vehicle that's going to get me there.' So how does pain get Carter to where he needs to go?

A lot of the research into how pain is adaptive, how people use it to help them get to where they need to go, is in the context of NSSI. As we discussed, this is problematic for a number of reasons, not the least being the stigma typically attached to NSSI. However, though there are specific physiological and psychological circumstances that apply to those who self-injure, the underlying mechanisms that turn physical sensation into effective emotional regulation hold up for everyone.

Dr Jill Hooley and Dr Matthew Nock from Harvard University, Dr Joseph Franklin of Florida State University, and Dr David Klonsky from the University of British Columbia are among the researchers approaching self-injury mechanically in an effort to understand how it functions in an individual's daily life. Collectively, their research suggests that many of those who self-injure do so to regulate negative affect and to promote positive affect, to reduce overactive emotions, and to manage or mitigate overwhelming feelings. Some models of the behaviour suggest that individuals who engage in the behaviour have low baseline levels of endogenous opioids and that these behaviours help them to restore opioid homeostasis.

There is no question that NSSI does regulate mood: Klonsky's research shows that high-arousal negative affective states such as anxiety, frustration, anger, tension and agitation all decrease following self-injury, while high-arousal negative affective states such as feeling calm and relieved increase. But as to why self-injury appears to help people reduce negative affect, the theories abound. It could be that NSSI offers a distraction or alternative to more upsetting situations in their environment or in their own mind. As Australian pain researcher Brock Bastian told us, 'Pain is just an excellent distraction. If I'm feeling depressed and I see a tornado coming towards me, I'm not going to be depressed at that time, I'm going to be worried about the tornado coming towards me.'

Because of how critical monitoring cell damage is to our survival, incoming nociceptive signals are prioritised within our complex allostatic-interoceptive network. That means that other signals are muted – including, for example, the signals underpinning troubling mental activities such as rumination. Rumination, in the clinical sense, is the rechewing of thoughts over and over and over, thoughts that you just can't seem to swallow; it's also a transdiagnostic feature of many psychiatric disorders like depression, anxiety, PTSD and ADHD, meaning that it often accompanies them. Rumination interferes with effective treatment options: mindfulness activities, reframing and reappraisal all require sustained cognitive control or the ability to focus. This becomes difficult or impossible for those who have intrusive thoughts that keep pushing them off the track.

But pain reliably captures attention; a 2019 mini meta-analysis showed that subjects engaged in less rumination after painful challenges, as measured by a thought-listing task and a self-reported rumination questionnaire. Meanwhile, imaging studies also show that pain competes with networks implicated in rumination and negative thinking – with attention and focus shifted to the body, there is no room for rumination. In fact, studies found that among those who self-injure, participants that also reported high levels of self-criticism experienced more relief compared to those who didn't.

This ability to capture our neural attention, so to speak, might also be why Doukas's subjects reached for electric shocks over reframing as a coping mechanism in response to the disturbing images. When we're already experiencing intense emotions, deploying regulation skills that require cognitive functioning and flexibility often fail – they're simply too resource-intensive to work.

Another part of why NSSI works, Bastian said, is that relief of pain is itself a positive emotional state. This is what's called 'pain offset relief' and it describes the lift in mood immediately after a painful experience. Most of the research on pain offset relief has been conducted in a lab environment with experiments that begin by introducing an aversive stimulus, such as super-loud annoying noises or electric shocks. The findings typically fall in line with the idea that pain offset relief is a result of both reduction of negative affect, usually by removing the aversive stimulus, and increase in positive affect via engaging the neural reward network; all of this is facilitated by endogenous opioids. Research suggests that because nociception and emotional distress share a great deal of overlapping neural circuitry, that relief of one kind of pain, say, pulling the knife from the skin, might also increase neurochemical activity that relieves the other. 'In short, the removal of emotional pain is difficult, but the removal of self-induced physical pain is simple and easy to control,' Hooley and Franklin wrote in a 2017 article in *Clinical Psychological Science* on NSSI.

Why people choose to pursue pain instead of another avenue for emotional relief is another question, and clues might be found in how nociceptive activation impacts our autonomic nervous system functioning, specifically the balance of sympathetic and parasympathetic activity. Remember, the sympathetic nervous system (SNS) is the 'get up and go' side of the autonomic nervous system, while the parasympathetic side (PNS) is the 'rest and digest'. Studies show that those who self-injure display increased sympathetic nervous system activity and decreased parasympathetic activity in response to stressors, an indication that their autonomic nervous system is putting the pedal to the metal (high SNS) with some faulty brakes (low PNS). Other research found that in people who self-injured, cortisol levels were higher than healthy controls when measured in the morning – suggesting they were already stressed when they woke up, possibly in expectation of strain.

Dysregulation in these autonomic systems may also explain the evidence showing that people who self-injure have a harder time recovering following highly intense interpersonal experiences, like getting into a fight with a parent, feeling rejected by classmates or loss of a loved one. If this is the case, then choosing to cut or injure oneself might be working as a way to jump-start the CNS when autonomic systems fail to turn over. And according to those who self-injure, it works. Moreover, because it works, the user keeps doing it.

This is the problem because, although self-injury may be helpful in the short term, the long-term consequences are anything but, including infection and permanent tissue damage. The correlations are equally concerning, including positive associations with a sense of shame, guilt and anger, and risk of substance abuse. But we can't say to what degree NSSI is the cause of versus the sought-out solution to emotional pain. It's also likely that part of the shame is tied to the stigma of NSSI, which can overshadow the person and the struggles behind the behaviour. And it isn't just parents, teachers or friends who have a hard time seeing past the injury. Academics and mental health professionals promote further stigma with the language they choose. Literature that makes frequent reference to how 'alarming' the behaviour is and the use of phrases like 'self-mutilation' and 'deliberate destruction of bodily tissue' can further ostracise a population that needs more support.

We know that pain, nociception, can work to decrease negative affect and increase positive affect, but we also know that hurting yourself to feel better in dangerous ways is a short-term gain, long-term loss. The question is: how can we use what we know in healthy ways that promote healing but not harm? With practice and training, this is the needle Aiden Carter learned to thread.

Fakir Musafar died in August 2018 from lung cancer at the age of 87. Carter was left feeling numb. In the previous two years, he had lost eight friends or family members, including his grandmother, and had had to put his dog to sleep. With each loss, he mourned, eventually coming to believe the universe – or something – was testing him. 'I was angry at it, whatever that might be,' he told us. Carter was left feeling numb. With the support of his friends, he chose to process through a flesh hook suspension.

Flesh hook suspension is a practice with ancient roots – ancient Hindus as far back as 5,000 years ago practised it as a kind of spiritual penance, and some still do at the festival Thaipusam. It's a powerful demonstration of just how strong our skin is but, to be honest, it is hard to watch: large sterile hooks are pierced through the skin. From there, a practitioner can go on to practise hook pulls or a full body suspension. There are many ways to go about it, different parts of the body and numbers of hooks from which to suspend, but safety and skill, as you can imagine, are critical – there's always a risk of infection or nerve and tendon damage if the hook is placed incorrectly. Some parts of the body are more conducive to suspension than others, Carter said: for example, on the back, right below the ribs, is the equivalent of the scruff of a cat's neck.

The most intense sensation happens the moment the feet leave the ground, and the full weight of the body is transferred to however many hooks through the skin. Carter said this is the most difficult part, especially for beginners; he's learned there's no use dragging it out: 'Whether it takes 20 minutes or five minutes to get off the ground, it's going to feel the same.' This is the moment of confrontation with the self.

But despite what it looks like, the dominant sensation of flesh hook suspension isn't, practitioners say, pain. Similar to endurance sports, many body stress rituals are independent meditations or performances, meaning you are the one pushing forward through the intensity with full awareness and agency. Carter explains this to beginners he is facilitating, 'You have to confront that sensation on your own. I can't submit for you; you have to be the one.' It has to be a choice, like standing at the edge of a high dive platform, looking down into the water and knowing you will be OK – or at least comfortable with the risk that maybe you won't be. In this moment, the mind surrenders to itself in the body, Carter says. 'When you relax past the perception of the pain, the pain drops away and becomes sensation.' What happens next varies, from transcendence to euphoria to delight to, as suspension.org warns, disappointment when none of those things happen.

In the wake of seemingly insurmountable loss, Carter sought healing in intense sensation, in the path that Musafar had taught him to walk. Surrounded by a cluster of trusted friends, he shouted and cried out, swung and twirled on the hooks. He was only suspended for 10 or 15 minutes, he said, but it felt like hours. His friends lowered him down and gently, methodically, removed the hooks, and cleaned the pulls left behind. Carter describes this suspension as the most intense he can remember. 'My experience was this range of emotion, of anger, aggression, mourning, and then at the end of it playing and joy,' he recalled. It was the opposite of numb.

What was an adaptive but dangerous discovery in his youth is now a risk-aware, conscious and intentional path towards healing.* And just as when he was young, he says, 'As I feel those hook pulls heal, I'm healing the emotion as well.' Over the next two weeks, the skin where he's been pierced is sensitive; each time he cleans them, each accidental bump is an invitation to reflect on his grief, to remember cherished moments, to heal.

If we were to categorise Carter's activities – and the activities of thousands of people who use ritual body stress for healing and spiritual worship – as clinical, as NSSI, then we would miss a lot. Carter uses body ritual and piercing to process and manage grief in a way that makes sense to him. Musafar taught him that wherever you are, whether you're actively practising body play or ritual or buying groceries, 'anything you do, you feel something'. This interoceptive skill underpins how he is able to turn pain – emotional, physical – into the less inflammatory intense sensation.

Carter learned to tap into the relief found in his own skin in a way that offers him healing without lasting harm. But hanging from some strategically placed hooks pierced through our skin is not for everyone, and we're certainly not advocating any kind of self-harm. However, Carter's story and experience remind us that emotional relief and even joy might be no further away than our own skin, in a body full of sensations waiting to be discovered. We don't have to resort to piercing – or any form of cell damage – to access the emotional regulation benefits that come with some somatosensory stimulation. When the volume on our emotions starts to get a little too loud, when our inner voice seems impossible to mute, tap into those interoceptive skills and feel your body. Literally.

One way we've already mentioned is simply going for a run. But it doesn't have to be running – all types of exercise have long been known and promoted as an effective aid in treating depression and anxiety and any number of challenges our mind sees fit to throw our

* Or just for fun: Carter shared clips with Margee of him and his friends smiling and laughing, zooming down a zip line on hooks through their own skin. As you do.

way. It's not just the endorphins – though those help – it's also because exercise forces our nervous system to reprioritise resources. The muscles that are 'feeling the burn' can scream louder than your intrusive negative thoughts. That's worth remembering when we're in emotional discomfort. Trading an emotional discomfort for a physical one helps us to if not always regain control of our experience, at least get out of our own heads for a little while.

This isn't just anecdotal or informed by an inspirational quote on Instagram. Studies have shown that intense tactile stimulation works just as well as nociceptive activation in disrupting thoughts and tasks, and keeping people from resuming those activities. The reprioritising of neural function is the basic theory behind using somatosensory stimulation as a means of emotional regulation and the inspiration behind cognitive neuroscientist Greg Siegle's 'emotional prosthetics' research. Full disclosure: Siegle is one of Ashley Doukas's mentors and has worked with Margee at the University of Pittsburgh for the past four years in understanding how – and why – people 'voluntarily engage with high arousal negative stimuli' (or, to put it simply, 'why we like being scared'). The thread tying all this work together is finding ways to capitalise on the power of sensory input to change autonomic functioning and ultimately how we feel.

In one of Siegle's experiments, participants in the test group received random tactile vibrations on their wrists or chests as they were exposed to a cognitive stressor exercise in which they have to add up a series of numbers as they appear on screen. The results were promising, with some participants 'experiencing profound reductions in subjective and physiological indicators of stress along with increases in behavioural performance.' These findings 'provide some of the first evidence that tactile vibratory stimuli can reliably alter subjective experience, parasympathetic activity, and performance on stressful cognitive tasks in tandem.' Beyond offering quick, non-pharmacological aid in the moment, this study supports further exploration of how tactile vibration can help those with conditions linked to dysregulation of the autonomic nervous system, including chronic pain, fibromyalgia, anxiety, PTSD and IBS.

This work is not only novel, it's desperately needed. Among the conditions listed above, the challenge of reducing chronic stress is common to all. The drug therapies available, mostly antidepressants, are only effective for a minority of people and come with serious side effects. Psychotherapy is time- and resource-intensive; many alternative therapies, such as yoga, simply aren't affordable or accessible. And critically, none of these approaches work as in-the-moment aids when stress goes off the rails. Let's face it, your co-workers would probably not tolerate multiple 20-minute meditation breaks a day. Using tactile vibrations to increase parasympathetic activity and bring down stress is affordable, accessible, easy to DIY, portable and discreet.

It's also a lot more practical, not to mention socially acceptable, than a flesh hook suspension.

Put a pin in it – using nociception *for* nociception

If some pain can mitigate emotional stress, couldn't it also mitigate somatic pain as well? The short answer is yes. We know this – we actually do have a framework for understanding that some pain can mitigate or even extinguish other kinds of pain. The mechanisms by which pain helps people manage emotional pain, when it does, are similar to the mechanisms by which it helps people manage somatic pain as well. And if it's both at once, well, that's even better.

Electrical stimulation, for example, has a long history of therapeutic uses, long before Ashley Doukas studied it in the lab. Electricity was used by the ancient Egyptians, Greeks and Romans as a remedy for rheumatoid arthritis, headache and other forms of nagging pain. Sufferers would expose the part of their body in pain to live electricity-generating fish, such as the Nile catfish or the torpedo ray, in the hope of inhibiting that pain. There is also evidence that people as far back as the third century knew how to make batteries – in the 1930s, archaeologists discovered an earthenware jar containing a copper cylinder holding an iron rod, sealed in asphalt near what is now Baghdad. Further analysis revealed signs of acidic corrosion. An American engineer made a replica of the jar and filled it with acidic grape juice; the live battery produced about 2 volts of electricity. Exactly what batteries in the ancient world were used for is uncertain – some theories suggest electroplating – but others have suggested they might have had a medical function.

Benjamin Franklin, the American polymath and politician, was particularly interested in electricity for therapeutic purposes. In the 1750s, he used a frictional electrostatic generator to treat people suffering from paralysis with both low-level continuous electricity and electric shocks, although with minimal effect. 'Franklinisation', or electric bath therapy, remained popular with quacks, but Luigi Galvani's discovery in 1780 that an electric current applied to the muscles of a dead frog's leg caused the leg to twitch sparked more rigorous scientific interest.* Throughout the 19th century, electrotherapy was used in diverse ways to treat conditions of intractable pain, disease and mental illness; electrical pioneer Nikola Tesla ignited a fad for what came to be called 'violet ray machines' when he introduced a device that allowed electrical currents to flow through the body at the World's Columbian Exposition in 1893.

The mechanisms behind how electricity worked therapeutically were variously interpreted, but the basic instinct was the same: a literal shock to the system. Big problems arose, of course, when electroshock therapy was used on people who were not asked or able to consent to the treatment. However, evidence bolsters the claim that electricity can be useful in a variety of ailments. Now we can and do employ electrical stimulation to counter pain all the time, using transcutaneous electrical nerve stimulation, or TENS. One theory as to why TENS units offer pain relief is that all the noisy afferent signals being sent up the line initiate descending pain-inhibiting activity, increasing the circulation of our endogenous painkillers. Descending pain modulation, then, is essentially pressing mute on nociceptive signals from the periphery. This approach is supported by evidence showing TENS units are most helpful when the intensity is turned up beyond gentle vibrations, meaning TENS units work best by treating pain with pain.

Nature also offers up an opportunity to treat pain with pain. For joint pain, backaches, arthritis and even neuralgia, capsaicin – the ingredient in chilli peppers that makes them hot – is widely popular.

---

* Galvani's discoveries, and the experiments of the many who followed, also inspired Mary Shelley, some 35 years later, to create one of the most enduring bogeymen in fiction: Victor Frankenstein's monstrous scientific hubris.

Scientists haven't figured out exactly how capsaicin cream works, but the thinking is that it has something to do with changes in the pain messenger called substance P. While capsaicin initially increases the release of the messenger, according to Harvard Health, 'After several applications of capsaicin, local stores of substance P (and possibly other chemical pain messengers) become depleted, and the nerve fibers in that area transmit fewer pain signals.' The real benefits of capsaicin only come with repeated use, and the suggested dosing is to apply over any painful areas four times a day. Results comparing it to placebo show it offers moderate relief, and adding some anti-inflammatories might increase its effectiveness. Still, with few side effects (as long as you use as directed), it's worth a try.

But TENS units and capsaicin aren't the only way that we use pain to treat pain. A hundred years ago, a procedure that involves inserting fine sterile needles into the skin all over your body and then maybe even adding some pressure, heat or electrical stimulation probably wouldn't have been thought of in Western cultures as anything except physical torture. And yet, now millions of Americans have tried acupuncture for everything from chronic pain and fibromyalgia to allergies and acne; according to a National Health Interview Survey conducted by the US Census Bureau that examined acupuncture trends between 2002 and 2007, the number of acupuncture patients in the US rose from 8.19 million at the start to more than 14 million at the close. And though the UK reportedly lags behind other European countries in its acceptance and promotion of acupuncture, when Linda was pregnant with her second child and vomiting rather more than she liked, her NHS-based antenatal care hospital offered her acupuncture as a treatment option; her own mother definitely wouldn't have had that option.*

Acupuncture is a practice in traditional Chinese medicine (TCM) that involves placing very fine needles into the skin at specific locations, called acupoints, to improve the flow of qi or chi, the circulating energy that, in TCM, animates all living beings. Manipulating this flow with the intention of restoring balance is meant to treat a variety

* She took it, but unfortunately the only thing that cured the nausea and vomiting was having the baby.

of ailments, including pain. Acupuncture also has ancient roots, although needling was by no means the most popular or considered the most effective of TCMs up through the 18th and 19th centuries. Meanwhile, use of acupuncture by 19th-century American and European physicians wasn't exactly common but it was fairly widespread. By the 1850s, however, physicians began to discard acupuncture when they couldn't determine the underlying mechanisms behind it. This was also the case in China. In fact, in 1822, the Chinese government tried to ban the practice altogether as part of efforts to modernise medicine, and by 1911 it was no longer part of the Imperial Medical Academy exam.

Change in opinion came relatively quickly, however, when nationalism catapulted acupuncture into the Chinese mainstream: after the Communist Party takeover in 1949, Chairman Mao Zedong began promoting acupuncture as an affordable, pragmatic way to offer healthcare to the masses, most of whom lived in areas without sufficient services. Enthusiasm outside of China for the therapy grew in the early 1970s: James Reston, an American journalist covering US President Richard Nixon's planned visit to China, developed appendicitis during his visit in 1971. His surgery was very much in line with Western surgical practices of the time; however, his post-surgical pain was treated with acupuncture by a practitioner who had practised for years on his own body. ('It is better to wound yourself a thousand times than to do a single harm to another person,' he told Reston.) Reston wrote about his experience for the *New York Times*, landing on the front page and sparking huge national interest. Acupuncture, it seemed, offered a way to treat pain and other ailments outside of the Cartesian, biomedical rubric that had left so many people disappointed.

Acupuncture was soon adopted by counter-culture health hippies and the growing alternative medicine movement. At the same time, that US doctors were now offering up scientific explanations as to why acupuncture effectively decreased pain made the practice even more appealing to a culture steeped in biomedical primacy but struggling with it. In 1995 the FDA recognised acupuncture needles as medical tools; in 1997 the NIH acknowledged acupuncture as an effective alternative treatment for a variety of health concerns.

Leaving aside the concept of qi – and most studies seem to – dozens of research teams have trialled acupuncture in many different conditions. A series of trials conducted by German researchers compared acupuncture treatments to standard medical care for patients with chronic lower back pain, finding that the treatment effect of acupuncture was 'significantly superior' to standard medical care, and that lower back pain improved after acupuncture treatment for at least six months. Another German trial examined acupuncture's effectiveness for people dealing with chronic shoulder pain and found that acupuncture improved shoulder movement and reported pain much better than standard medical care, leading them to suggest that it is an effective treatment. It's also worth noting that these trials were paid for by German health insurance agencies, and regardless of underlying motivations, the trials did result in insurance companies covering acupuncture as a viable treatment for specific conditions. *Cochrane Review*, our favourite one-stop shop for scientific validity, confirmed the effectiveness of acupuncture for sufferers of episodic or chronic tension headaches, and found that acupuncture reduced the frequency of migraine attacks in robust studies.

So far, perhaps unsurprisingly, much of the work on acupuncture has been in how it can help somatic pain conditions; a critical mass of random controlled trials measuring acupuncture for treatment of mental health related issues still has some way to go. But early evidence is encouraging. One review of 39 meta-analyses and 59 random controlled trials published in 2019 in the journal *Complementary Therapies in Medicine* found that acupuncture reduced the severity of depression slightly, but significantly more compared to treatment as usual, antidepressants, sham/control acupuncture, and no treatment. Meanwhile, the American military is cautiously exploring acupuncture interventions for service members dealing with stress; according to a feasibility study into rolling out such a programme, 'Preliminary findings suggest that [standard stress acupuncture] may be useful in improving energy/fatigue, social functioning, and perceived stress of service members.'

Not all pain conditions are the same, of course, so acupuncture isn't effective in all of them. According to *Cochrane*, for example, acupuncture offers little to no benefit to people suffering from rheumatoid arthritis or fibromyalgia. Still, as the evidence of acupuncture's effectiveness in certain conditions piled up, the focus has shifted to teasing out just how acupuncture works.

One big theory is that it's because it kind of hurts. The experience of 'de qi', which is the distinct stimulation that the needles offer, is described as an aching, soreness, numbness, tingling or heaviness and it is an expected part of acupuncture treatment. Acupuncture might not 'work' because it taps into energy lines or shifts qi – it might work because sticking a needle in your body initiates a nociceptive sensation that forces us to refocus on that pain instead of other pain, and that still produces a waterfall of endogenous painkillers.* The experience might also stimulate changes in neural architecture: electro-acupuncture, which combines TENS stimulation with needle therapy, has shown evidence of changes in the brain linked to symptom improvement; in a study involving carpal tunnel syndrome sufferers, fMRI revealed changes in the somatosensory cortex linked to improving functioning in the median nerve, the nerve that exits the cervical spine and runs down the arm to the hand and that is compressed in carpal tunnel syndrome.

However, one of the big reasons that acupuncture works might not be because of any specific changes the therapy itself is inducing but rather changes that our own systems are generating. One of the other findings from the German acupuncture trials was that sham acupuncture – needles inserted in places that shouldn't have any effect or placed extremely shallowly, or participants being tapped with needles that did not pierce the skin at all – worked almost as well as real acupuncture and better than standard medical care. Other studies have had similar results. For example, a 2009 meta-analysis of 22 trials with 4,419 patients found acupuncture at least as effective for managing

* At her own wedding reception, between the first dance and the cake-cutting, Linda's low-level migraine turned into a high-level migraine. Luckily, one of her bridesmaids was prepared. A licensed acupuncturist, she carried needles with her at all times. So Linda sat at the bar of the venue, in her wedding dress, while her friend applied the needles to her hands and feet. Linda experienced a marked reduction in symptoms, including lower levels of pain and visual disturbances, while the needles were in, but sadly the effect didn't last after the needles were removed. Still, the cake was good.

migraine as prophylactic drugs and with fewer side effects; so was sham acupuncture. What this demonstrates is that relief results not only from what is 'real', but also from what we perceive to be true. It demonstrates the power of one of the most useful and under-studied tools in the caregiver's bag: placebo.

The power of placebo

Elisha Perkins could be one of medicine's early heroes, a man who relieved the pain of thousands of people living with chronic conditions such as rheumatism and neuralgia. Or he could be one of the greatest charlatans of the late 18th century. He's probably both.

Perkins, born 16 January 1741, was a handsome Connecticut doctor who stitched up musket holes and set bones during the American Revolution. After the war, he started his own private hospital for wealthy clients. During a tooth extraction, he noticed that touching his metal tool to the patient's inflamed gums temporarily took away the pain. This gave him an idea and he got to work. In 1796, he patented his miracle pain cure: two thin, 3-inch rods made of a 'special metal alloy'. He called the rods 'Tractors' and promised that placing the Tractors on the painful points of the body for 20 minutes would rid the body of the 'toxic electrical fluid' causing the pain. Patent in hand, Perkins took his Tractors to market in Philadelphia, where Congress was then meeting. Perkins somehow got his Tractors in the hands of some very powerful people – George Washington bought a set, as did Supreme Court Chief Justice Oliver Ellsworth – and the new treatment took off.

Perkins claimed he'd cured more than 5,000 people of ailments as varied as gout and tooth pain; his claims were backed up by professors, physicians, clergymen. Even the royal physicians to the court of Denmark tested his Tractors and declared them legit; they called the treatment 'Perkinism'. According to the New England Historical Society, 'Perkins had no shortage of partners. A man in Virginia gave up his plantation and invested everything in Metallic Tractors. Cases were reported of people selling their horse or carriage to buy a Metallic Tractor. People soon began using the Perkins Metallic Tractors on their animals – with great success, they felt.'

Meanwhile, Perkins was condemned by fellow physicians, not entirely because his Tractors were utter quackery; the fact that patients could and did use these treatments without the aid of a physician was a threat to their livelihoods. In 1797, Perkins was booted out of the Connecticut Medical Society. He wasn't bothered, though: his Tractors had made him rich and respected, albeit not by his fellow doctors, and he felt he'd done humanity a great service. Sadly, he didn't live long enough to enjoy it. In 1799, he'd formulated an antiseptic involving vinegar and salt that he said could cure yellow fever. He tested the elixir during an outbreak in New York City, where he died of … yellow fever.

Nevertheless, his Tractors remained popular after his death – despite compelling evidence that they were complete frauds. In 1800, Dr John Haygarth, a methodical and curious London doctor, swapped the special Perkin Tractors – which were really just steel and brass – for wooden ones painted to look exactly like the real ones. He found that in treating patients suffering from rheumatism, the fake Tractors worked just as well. Haygarth's article, published to mild interest, declared that the 'whole effect undoubtedly depends upon the impression which can be made upon the patient's Imagination.'

The thing was, they worked. Though they certainly didn't work the way Perkins claimed they did, nor the way the people who used them thought they did, they still worked. Why did they work? Because of an incredible superpower that we all possess – placebo.

Although it might not have been thought of as such, placebo has been used to effectively treat ailments for millennia. Researcher Luana Colloca from the University of Maryland noted in the *Annual Review of Pharmacological Toxicology*, 'An estimated 5,000 ancient remedies with over 16,000 different prescriptions, from Gascoyne's powder (i.e., coral, crabs' eyes and claws, amber, bezoar, and pearls) to bezoar stones and animal gallstones were initially used to please or placate and not for specific clinical effects.' In fact, explicitly placebo remedies appeared in the first London Pharmacopoeia issued by the Royal College of physicians in 1618.

Modern science confirms the power of placebo. In addition to the studies showing the effectiveness of sham acupuncture, significant effects have been documented in improving symptoms from a wide range of disorders including Parkinson's, ADHD, anxiety, social phobia, depression, addiction, asthma and even some immune diseases. Evidence points towards pain relief resulting, at least in part, from our predictive processing; our mental simulation calls into action our own pain-regulating tools including endogenous painkillers. In fact, studies across disciplines demonstrate, as one put it, 'that placebo effects engage various neurobiological and physiological mechanisms, including the endogenous opioid, endocannabinoid, oxytocin, vasopressin, and dopamine systems.'

If predictive processing can generate positive, healthful effects, then it can also generate the opposite; this is the nocebo effect we've mentioned. A study – 'Placebo and Nocebo Effects: The Advantage of Measuring Expectations and Psychological Factors' – published in the journal *Frontiers in Psychology* in 2017 demonstrates how it works. In this study, participants were first conditioned to expect that a series of colours – red, yellow and green – delivered high-, medium- and low-thermal stimuli via a thermode attached to their forearm. During the experimental trials, participants were shown one of the colours for four seconds, and quickly asked what level of pain they expected, followed by 10 seconds of the thermal stimulation, and then directly afterwards participants were asked to rate their perceived pain intensity. Critically, the thermal stimulus delivered was always the same medium intensity, but that didn't matter; participants' perceived pain intensity matched their reported expectations based on the previous conditioning.

The intensity of the placebo and nocebo effect varies in some predictable ways; the worse we expect to feel, well, the worse we feel. This study also examined how personality traits matched with experience. Severity of anxiety and higher scores for depression, irritability, guilt, feelings of being punished and fatigue were significantly related to lower placebo effects (less pain relief), and greater discomfort with sensations related to anxiety, such as a racing heart, high physiological suggestibility (for example, feeling hurt after someone says, 'That must have hurt') and pain catastrophising were significantly related to greater nocebo effects (more pain). Those in the study who reported less fear of pain experienced greater placebo effects (more pain relief).

But the really fascinating thing is that placebo – and nocebo – does not require deception to work. A 2014 study by Ted Kaptchuk, a placebo effect researcher with Harvard-affiliated Beth Israel Deaconess Medical Center, found that sugar pills in a bottle explicitly labelled 'Placebo' were still 50 per cent effective compared to the real migraine medication. This illustrates how placebo is not just about 'tricking' the mind, but employing the body to do the work we wanted the pills to do. 'Even if they know it's not medicine,' said Kaptchuk, 'the action itself can stimulate the brain into thinking the body is being healed.'

Right now, mentally walk yourself through the process of taking a pill. You are holding the prescription bottle in your hand, using your fingers to push down and turn. There's the sound of shaking out a pill into your open palm; you toss it into your mouth, tasting the bitter flavour of the pill coating as you wash it down with a glass of water. When practiced enough times, all of the sensory signals and motor coordination – along with whatever chemical, biological changes result from the pill – are encoded into a predictive model. When enough signals show up together, we run the model – and get some of the relief without the stuff that supposedly brings the relief. It may not be as powerful as the real deal, but it helps.

Again, this is the power of predictive processing, of creating a mental simulation that generates real neurochemical changes – prediction and simulation not only shapes perception but *is* perception. As Kaptchuk told *Men's Health Watch*, 'The placebo effect is more than positive thinking – believing a treatment or procedure will work. It's about creating a stronger connection between the brain and body and how they work together.'

What's also increasingly evident is that non-pharmacological placebo interventions, such as sham surgery and sham acupuncture, might carry more clinical weight and be in a different class altogether. For example, a team from Technische Universität München in Germany found that sham placebos, like sham surgery and acupuncture, are linked to stronger, though non-specific effects compared to oral pharmacological placebos. Further research suggests that context plays a big role in this difference – specifically, pain in the context of treatment is processed differently on a neurobiological level as compared to pain in other contexts. In a 2015 study, one group was told they were being given acupuncture; the other group was told a needle would be inserted and it may be painful. Imaging data was collected on both groups as they underwent their procedures; analysis revealed that the group told they would receive the acupuncture treatment had increased activity in reward circuity and decreased activity in pain related networks. This work suggests that networks for pain and placebo work in overlapping but different ways thanks to positive expectations – therapeutic pain becomes pain with a purpose, pain with hope.

It was with you all along

Placebo, like everything we've talked about in this chapter, suggests some really powerful potential treatments that would cost virtually nothing to harness.* The uses of pain, even the suggestion of pain, go far beyond just a teaching tool – you might already be using pain in this way without even realising it.

Finding novel treatments for pain, both emotional and physical, is more important than ever, as both chronic pain and chronic stress are on the rise all over the world. In addition to estimating that more than 20 per cent of the global population suffers from chronic pain, the World Health Organization found that depression affects as much as 4 per cent of the world's population, a cohort that is growing. Independently, these figures are worrying; together, they compound already dramatic effects on health and well-being, not to mention the knock-on effects on everything from economy to education. Meanwhile, our treatments for both are largely reliant on pharmacological interventions with equally spotty track records. We've already discussed how poorly most drugs work for pain; according to the latest update from the NIH in 2017, after taking placebo effects into account, antidepressants help only 20 people out of 100 who struggle with moderate or severe depression (for those 20 people, however, they are a lifeline).

What if the cure for pain could be found in our own bodies? Aiden Carter found it. The millions of athletes who rely on their daily

* Of course, how do you go about making appointments and paying for (or getting health insurance to pay for) a treatment everyone knows is a sham?

confrontation with pain have also found it. Researchers who are looking past the stigma of NSSI and into the science of how some kinds of physical sensation can be used to manipulate other kinds of sensations have found it too. Acupuncture users found it. Even Elisha Perkins, with his incredible Tractors, found it, though he probably didn't mean to.

We know that emotional pain stings just as hard, and sometimes harder, than damage to our physical body. In those instances, of loss and grief, of feeling too much or not feeling enough, turning up the dial on physical sensation in safe and thoughtful ways can be a pathway to relief. Similarly, meeting somatic pain – especially somatic pain that is under your control – with more somatic pain and a little bit of intention works in the same way. That we can harness these effects without tissue damage is a testament to the power of our very efficient predictive processing systems, driven by our allostatic-interoceptive networks. If that's not a miracle cure, nothing is.

CHAPTER SEVEN

Hurts So Good: When Pain is Pleasure

On a warm Saturday night in May at a small Pittsburgh theatre, around 300 people pressed shoulder to shoulder to watch a woman performing a burlesque routine on stage. That in itself wasn't extraordinary; burlesque was mainstream even before Cher and Christina Aguilera made a movie about it. But tonight's performance wasn't about sex, not exactly.

Macabre Noir is petite – just 5 foot 3 inches – and voluptuous. Armed with two huge black feathered fans, she's dressed in an outfit straight from the wardrobe closet of the Moulin Rouge – a jewelled black and red satin corset, arm-length red satin gloves – that makes her porcelain-white skin glow. As eerie carnival music plays, she slowly peels off her gloves, revealing her tattoos – a heart, a massive Lovecraftian cross, a portrait of Edgar Allen Poe, another of magician and mountaineer, 'the wickedest man in the world', Aleister Crowley. Her long hair is a fluorescent red shade called 'Vampyre', shaved clean on both sides, leaving her with a bright mane of hair down the centre. Her natural eyebrows have been replaced with pencil-thin high arches, perfectly matched to her hair colour. The pearls she wears are black; one is fixed in the piercing above her top lip, one under her bottom lip. Thick wooden spiral earrings loop through each of her ears.

The crowd is silent and mesmerised as they watch Macabre take off her clothes in a seamless dance. Eventually, she stands on stage in black high-waisted mini-shorts and a black satin bra with intricate beading that sparkles in the light. The music drops to a low thrum.

Looking out across the audience, her expression shifts from one of seduction to mischief. She walks to the side of the stage, picks up a black bucket and presents it to the audience. She lifts the bucket high into the air and dumps the contents out in a narrow path across the front of the stage. The tinkle of broken glass – pieces of thick green wine bottles, brown beer bottles, clear soda bottles – echoes through the quiet theatre. In the bright spotlight, it looks like a glittering stream. A single grey, doomed balloon slowly floats down from the rafters above centre stage. It lands lightly on a jagged piece of glass and pops; some in the audience let out startled screams.

Barefoot, Macabre steps onto the sparkling path. There is a collective gasp of breath: this is what we – the punkish Goths, the curious hipsters, the college kids who look like they might be lost – came to see.

She is tentative at first, holding her arms out from her sides for balance, steadying herself as she plants one foot and then another on the broken glass. Never breaking eye contact with the crowd, her face contorts into expressions of pain and fear, a stark difference from her earlier confident and sultry strip-tease. She reaches the other side, and we erupt into applause, although we don't know whether we're clapping in celebration of her feat or because it is over.

Macabre turns, takes another crunching step onto the glass. It is not over. Halfway across, she stops. All signs of trepidation are gone as Macabre straightens. We've been played: with a devious grin, she jumps, landing with a crystalline crunch. As the dark carnival music picks up again, Macabre leaps, prances, dances on her river of broken glass. Some in the audience close their eyes, murmuring, 'I can't watch.' We gasp at each crunch, each time her pale, naked foot lands on the glittering jagged glass; each new move is met with waves of 'Oh, my gosh!' and 'Jesus Christ!' Finally, Macabre steps off the path, brushing her feet off and disappears off stage. She's back a second later, carrying a bar stool. She climbs and then, barely pausing for the effect, jumps off, bringing both feet down onto the twinkling shards. Small pieces of glass fly a few inches, displaced by the force.

Macabre grins widely, opening her arms and taking a bow. She turns to leave the path, but pauses. Someone whispers, 'Oh my god, I think she's bleeding, she's totally *bleeding*.' Macabre doesn't say anything. She shakes her left foot, her right foot, then bends down to brush the shards of glass from the bottom of her feet. She pulls a stubborn sliver from her heel, throws it into the pile behind her, then straightens, smiles and trots off stage as the crowd breaks into applause.

Macabre has been performing artistic feats of pain since she was 25. She's now in her early 30s and a frequent performer on the dark arts cabaret scene. Her first act involved attaching wings to her shoulder blades with surface piercings. This was an artistic evolution of something she used to do as an adolescent – then called Kelly Braden – in rural western Pennsylvania: sticking safety pins through the webbing between her fingers and holding them up to shock her classmates. 'I loved freaking them out, getting the "ewwws",' she said, chuckling. She was, she said, a 'weird, weird child and an even stranger teenager.'

Macabre only stopped the wing piercings after building up too much scar tissue on her back; it became, she said, too much work to break through that skin. Now her main act is dancing on a bed of broken glass and climbing a ladder made of blades.* The 'ewwws' – a mixture of pleasure, awe and fascinated revulsion – are part of what she thinks brings people to her shows. 'People have a morbid curiosity. It's the reason they go to haunted houses, the reason they watch horror movies. The appeal is that it's live, happening in front of them and it's real, it's not a movie, not a haunted house,' she said. 'It's very real and a huge draw to say, "I can't believe this person is doing this thing right in front of me."' That might explain why people come to watch her, but not why she does it. Superficially, she does it for the same reason most people do anything: because she likes it and because she can.

Challenging normal

Outside of her communities, people make a lot of negative assumptions about Macabre, mostly because of her appearance. But researchers have also made a lot of negative assumptions about people like Macabre and their atypical relationship with pain. Many researchers have tried to understand why people would enjoy painful or otherwise 'negative' experiences, and most have got it wrong – because they're asking the wrong questions.

---

* Glass-walking doesn't carry as much risk of tissue damage as it looks; this is largely owing to physics: Many uniform pieces of smallish glass will settle and shift under your feet as you're walking on them, allowing your weight to be better distributed across them, somewhat like walking across sand. You're not stepping on any one piece with enough force to puncture your foot; this is also why laying on a bed of nails is infinitely preferable to stepping on just one. The blade ladder, now, that's a bit riskier; balancing on the blade distributes weight across a larger area, not unlike the glass. But if you slip, you might find yourself losing a chunk of skin or worse.

As kids, we twirl in circles and fall to the ground, delighted by our dizziness. We push, pull, pinch, jump, caress and investigate the boundaries of sensations just to see what it feels like. But experience, culture and the opinions of others quickly teach us what is 'good' or 'bad' to do to our body, both in the company of others and in private. This internalised behaviour is steeped in emotional meaning: doing 'good' things to our body is supposed to make you feel good, bad things are supposed to make you feel bad. Pain is supposed to be one of the bad things – and sometimes, it certainly is – but the boundaries of what is good or bad to do to our own body are defined by the culture, religion and the time in which we live. Often those boundaries are far too narrow.

'There's a lot of people like us,' said Macabre. 'A lot of people hide it.' There's no doubt she is correct, and why they would hide it is exemplified in some of the research on enjoyment of supposedly 'bad' things. In a paper published in 2013, University of Pennsylvania psychologists Paul Rozin and co-authors coined the term 'benign masochism', also referred to as 'hedonic reversals', to describe the pleasure people found in 'painful' acts. The concept got a lot of press; however, though the article started some much-needed conversations, it has a lot of problems, not least of which is reinforcing the myth that experiences and our physical responses to them are innately good or bad.

In the study, participants rated how much they like 30 items that the authors defined as instances of benign masochism, including watching sad or scary movies, eating hot chilli peppers or aged blue cheese or drinking whisky. They found that 'people tend to like their physiological reactions to the innately negative experiences,' citing examples of enjoying eye-watering, sweating and heart-pounding. 'We explain these findings in terms of benign masochism, enjoyment of negative bodily reactions and feelings in the context of feeling safe, or pleasure at "mind over body",' they concluded. Benign masochism and hedonic reversal, as they describe, is enjoyment of 'bad' but not 'too bad' experiences.

But here's the problem: how are they defining bad? What is the line between 'benign' masochism and 'real' or even pathological masochism? For starters, the items on the list seem to be chosen almost at random; none of the items are 'innately negative' – just ask the millions of fans of horror movies, stinky cheese, whisky and spicy food. At the same time, the 'negative' physiological reactions to which they refer – the eye-watering, sweating and heart-pounding – accompany a host of activities that many people would characterise as 'good'. We can talk about the resources that these different reactions recruit in the body, how these stimuli change neurochemical signalling and what kinds of chemicals are involved. And we can speculate as to the cause and the function of these responses – for example, tears function to protect the eye or clear debris, but they are also a side effect of some drugs, hormonal changes and strong emotion. Yet we cannot say whether they are negative or positive, good or bad. The meaning is formed by the context and none of these physiological changes are experienced in a vacuum: if you find yourself enjoying the eye-watering and mouth-puckering of a handful of sour gummy worms, then that is a *positive* physiological reaction.

Making sense of why people like experiences described as benign masochism – or, really, most experiences that some would call negative or painful – is easier when viewed through a constructionist approach. The constructionist approach recognizes that context, including who you are with, your desires and motivations, along with previous experience and expectations goes into building meaning. We cannot tell someone they are misunderstanding pain as pleasure, because their pain *is* pleasure.

As Australian pain researcher Brock Bastian told us in an interview in 2017, many people do seek out pain, in a wide variety of contexts. Asking why they do it is sort of beside the point: 'I don't know that a great many people have a great understanding of why they do it ... they have an intuition of why they do it,' he said. It does something that they like and it's useful.* A much better question to ask, then, is what meaning people make of different experiences: 'The meaning of pain determines whether it's viewed negatively or positively,' said Bastian.

---

* One of the biggest challenges in researching pain? 'I thought I'd stumbled across something that was very easy to manipulate in the lab, but it turns out no one knows what it means,' he said, laughing. 'Pain is a dependent variable.'

Physical pain for Macabre Noir, she says, is 'a non-issue' – when it's under her control. 'I'm not going to fall down the stairs into a pile of broken glass and say "Hey, that felt great!"' she cautioned, laughing. But when Macabre chooses to engage in behaviour that the University of Pennsylvania psychologists would likely think goes beyond 'benign' masochism, she reports experiencing pleasure and even euphoria from pain in controlled settings. 'Your whole body feels amazing, strong, powerful,' she said.

Physiologically, there may be several mechanisms behind Macabre's positive affect, much of which we covered in the previous chapter. For example, she says it's not when the needle pierces her skin that she feels elation but rather when it's pulled out, in the relief of pain. 'I [feel] like I had four cups of coffee. Totally energised.' Some of this is likely to do with the endorphins released during painful experiences; some of it might have to do with pain offset relief, the elevation in mood after a painful experience is over. In fact, nociception can turn up the dial on a lot of our senses. In his research, Bastian found that pain increases sensitivity to gustatory input – flavours were perceived with more sensitivity and intensity,* suggesting this heightened pleasure is part of why pain offset relief works to improve mood.

Finding pleasure in the physical sensations is part of what makes this experience enjoyable for Macabre, but she also builds meaning, and finds strength, in the social context. Her ability to accept pain makes her feel capable, strong: 'Normal people don't want to, or can't do this sort of thing … It's this massive sense of accomplishment, it's a moment of invincibility.' Anyone who's run a marathon or done a Tough Mudder, for example, can understand that; certainly Rami Ibrahim, the Muay Thai champion, does. Macabre has also come to understand that for her, it's not just about that feeling of power – she likes feeling that powerful *in front of an audience*. 'I realised even though I enjoyed the piercing, I really enjoyed people's reaction to what I was doing,' she told us. She likes the attention, the command over those 'ewwws'.

Macabre's strength isn't only evident in her performance but in her commitment to being herself – knowing what she likes and

---

* If the next big thing in gastronomy is minor electric shocks, we want a cut of the profits.

owning it. It may sound like a quote on a Pinterest inspiration board, but self-expression and self-discovery can be a rebellious act. Yet it is one that can pay off in big ways; when we shed restrictive social taboos and ideas of what experiences *should* feel good or *should* feel bad, when we let sensation arrive unencumbered, we open up a world of opportunity.

Let's take an experience that many people hold up as the gold standard of pain: childbirth. Childbirth is the crucible that moulds mothers. It is damaging to the maternal body in ways that go far beyond cosmetic changes. There's no guarantee, even now, that both the mother and child will survive, and for thousands of years many didn't. The only pleasure that's meant to be found in childbirth is when it's over and you're holding the baby. And yet, some women report not just feeling good but actually orgasmic. As in, they had an orgasm during labour.

Debra Pascali-Bonaro, a doula and advocate for birthing mothers, is the director of the documentary film *Orgasmic Birth: The Best Kept Secret.*[*] Though she didn't herself have an orgasm during her labours – 'I would never want to have that as the performance standard, that would be awful,' she said – her experience was more joyous and less terrifying than the dominant narrative of childbirth would allow.[†] This led her to become a trained doula, where she witnessed the power of allowing women to make their own meaning of childbirth. And yes, she said, some women she attended did in fact experience orgasms.

'There is this orgasmic energy in birth and that does include some people who have birth-gasms,' she told us. 'But they kind of had it with shame – why did they experience pleasure when everyone else only ever talks about pain?'

---

[*] Linda here: talking to Pascali-Bonaro was such a comforting experience, it actually made me think 'what if' to another baby. And then I thought, 'You have lost your damn mind.'

[†] See *Knocked Up*, *Nine Months*, *Baby Mama*, *Bridget Jones's Baby*… literally any romantic comedy ever that depicts a woman in labour finds hilarity in the horror. It's funny but not because it's true. Bonus points for *Bridget Jones*, which managed to also get in a dig about non-pharmacological pain coping mechanisms: when Bridget's crunchier father option tells her to 'think away the pain' and remember her yoga, Bridget howls, 'Fuck yoga!'

Pain is dictated by context. And there is a lot of context around childbirth. It is a dense thicket of thorny emotions and expectations, judgements and individual traumas, needs and wants. But this context is also why finding pleasure in the pain – or rather, intense sensation – of childbirth is possible. The theory behind orgasmic birth is that anatomically all of the sensory 'pathways', so to speak, that are involved in sexual pleasure are also involved in the childbirth process. Leaving shame or guilt aside and allowing sensation to just exist, even reading it as pleasurable, can offer benefits including a more joyous childbirth experience, less fear and the potential for a rocking orgasm. And that in itself could offer analgesia: there is evidence that during pleasurable vaginal stimulation, women's threshold tolerance to nociceptive pain increases dramatically, by as much as 36 per cent. When women orgasmed, that tolerance shot up by nearly 75 per cent. 'Orgasm,' confirmed Pascali-Bonaro, 'is a huge pain reliever.'

How many women actually orgasm during childbirth is unclear. A 2013 study that appeared in the journal *Sexologies* found that just 0.3 per cent of women reported experiencing orgasm during childbirth. A 2016 survey from Positive Birth Movement and Channel Mum found the figure was closer to 6 per cent, although theirs was not exactly a representative sample – if you're already on a website about positive birth, the chances are greater that this would be a part of your experience. But many more mothers could have an orgasm during labour, or at least enjoy greater empowerment during the process, say advocates. Creating a space where birthing mothers feel protected, safe and respected is vital to reshaping the birthing process, a process that many women regard as traumatic and alienating. If they use that space to generate pleasure for themselves, whether through self-stimulation or by engaging in sexual activities with a consenting partner, then so be it.

Fakir Musafar said that there is no pain, only intense sensation. Neurochemically, biologically, this is true – pain is a meaning we ascribe to an experience. Pleasure and pain, then, do not exist in a binary, in opposition to one another. When we can experience pain not solely as the opposite of pleasure, we open ourselves up to some pretty incredible experiences. At the very least, we expand our emotional diversity, and learn more about what we do and don't like. And at the most, we discover a whole new world of pleasure.

A little slap and tickle

We would be utterly remiss if we didn't address the latex-clad elephant in the room – no conversation about pleasure and pain would be complete without talking about BDSM.

Obligatory disclaimer: a complete written exploration of BDSM requires a much, much bigger book than this. Several books, in fact, touching on history and culture, psychology and law, politics and religion. But what we're interested in is where this practice intersects with pain and, with that in mind, what it means, how it is experienced, how it is helpful and how it is harmful.

Today, BDSM is often used as a catch-all acronym for doing kinky stuff in the bedroom, but the letters and their configuration can be used to communicate a lot of information. The B is for bondage; the D can be used to refer to discipline or dominance; the S can be used to refer to submission or to sadism, and the M refers to masochism. While 'sadism' does come from the infamous Marquis de Sade – the French 'libertine' pornographic philosopher whose prolific texts detailed sexual gratification at the humiliation of and violence against women, men and children – the Marquis de Sade would not find a place in the BDSM community today: his activities were neither safe, sane nor consensual.* Sadism practised within the principles of the BDSM community occurs between willing adults, does not intend to cause trauma, to leave a

* In the years since his death in 1814, there have been numerous efforts to rehabilitate Sade's image, from feminist readings of his works that exonerate him for creating 'space' for women's sexual power to a proposed Sade line of Victoria's Secret lingerie. In the 2000 film *Quills*, starring Geoffrey Rush, Kate Winslet and Joaquin Phoenix, Rush's Sade is a grizzled charmer in the final years of his life, confined to an insane asylum because his work is too risqué for French society (witness the scene of virginal Kate giggling over his writing while hanging up the laundry). This, the film implies, is censorship, artistic freedom squashed because of zealous morality. If you have read *120 Days of Sodom* – and Linda has and does not care to do it again – you might disagree with this simplistic assessment. In real life, the Marquis de Sade was a sexual predator and serial rapist who grotesquely abused those of a lesser social rank. In 1774, for example, he trapped six young people in his château in Provence, where he tortured and raped them over six weeks. So maybe that glass of Sade wine is a bit inappropriate?

partner 'damaged' or to cause lasting distress. (Some self-identifying sadists might be aroused by non-consent – however, those who seek another's pain without their consent and further cross that line are crossing out of the BDSM community and into the criminal one.)

A lot of this very varied practice hinges on the enjoyment of nociceptive activation – whether that's through a light slap on the bottom, being tied up in restraints, or judiciously applied clamps – in the context of sexual arousal (but not always). We've already talked about how pain can promote positive affect and how pain offset relief can make gustatory pleasure that much more acute; a little nociception can also turn up the dial on other sensations – for example, a little whipping can make the gentle stroking that comes next feel all the more intense. We also know that sexual arousal alone can reduce pain, A 2012 study from researchers at the University of Münster found that sexual activity during a cluster headache or migraine might actually relieve symptoms: sexual activity offered moderate to complete relief of symptoms for around 60 per cent of migraine sufferers and 37 per cent of cluster headache sufferers. Some – though perhaps not enough – pain specialists even advise chronic pain sufferers to masturbate, on the speculation that the endorphins and other neurochemicals released during orgasm can offer endogenous pain relief.*

But the pleasure born of BDSM goes beyond inciting or reducing nociception. Adding nociceptive activation to sexual arousal is a bit like revving our sympathetic nervous system and parasympathetic nervous systems in complementary ways. Nociception amps up sympathetic nervous system activity, and that high arousal can be experienced as excitement, anticipation, and even a little fear (the fun-scary kind). Sexual arousal, on the other hand, involves the parasympathetic system and is associated with increasing the circulation of hormones, like oxytocin and vasopressin, linked to feeling good, connected and intimate. Skillfully engaging both systems at the same time can be a potent cocktail that amplifies, rather than counteracts, the effects of the other.

---

* In 2019, sex toy makers LELO wrote an open letter to the UK Department of Health, urging the NHS to have doctors prescribe masturbation to patients as part of an integrated wellness programme.

Research from University of Oxford's Dr Irene Tracey, one of the world's leading pain researchers, offers some insight into how this combination can intensify pleasure. In addition to being an expert on pain, Tracey is also an expert on the complicated relationship between dopamine and endogenous opioid signalling when it comes to regulating pain and pleasure. She and her team conducted multiple studies looking at how these two neurotransmitters impact each other and found that time and, not surprisingly, expectations are big factors.

Both pain and pleasure are associated with increasing endogenous opioids. When we're doing something we enjoy, opioids work to increase reward value, for example upping the pleasantness of sweet tastes. At the same time, opioids also decrease the intensity of pain; experiments where opioid signalling is blocked resulted in less interest in food and sex, *and* less pain relief. Dopamine plays a role in this whole process, working to either enhance the effects of opioids or block them. In situations when we're expecting reward – like sex – short bursts of dopamine can work to increase opioid signalling, which in turn decrease pain and increase 'pleasantness'. This dynamic process works as a moderator for motivation; the theory is that we essentially either 'turn down' pain so we can gain rewards essential for survival – such as food and sex – or 'turn down' pleasure when we really need to pause and tend to an injury. While this research was not conducted in the context of BDSM, it suggests that sexual arousal can dampen the pain but doesn't dampen the delivery of the endogenous opioids that come with pain. So the pain is less intense, while the pleasure of sexual arousal feels *more* pleasurable.

But the pleasure, really the enjoyment, of BDSM goes beyond just pushing the right buttons on our nervous system; for many, BDSM, practised privately or in community settings, offers physical pleasure, deep social connection and powerful personal growth. Which is why understanding BDSM – especially outside of the context of 'deviance' – is important for all of us, not just those of us who might enjoy a little spanking every once in a while.

First, some history. While there are many manifestations of the sex and pain combination, historical evidence shows there is a long history of the 'among consenting adults' variety. Written records of what would today be considered BDSM practices can be found in the earliest recorded texts, the Sumerian cuneiform, and the sexual rites of devotees of the ancient Mesopotamian goddess Inanna. The Kama Sutra, famous for its depictions of acrobatic sex positions, also includes lessons in eliciting joyful cries of pain from a partner via pinching and biting different parts of the body. That the ancient Hindu text is actually a manual in the art of living a meaningful life underscores how normative a good sex life is and the ways that pain informs that.

Jumping ahead a few centuries, modern BDSM groups in the US and the UK organised predominantly through gay communities, finding partnership and support among other marginalised groups. Being in any niche community prior to the age of the internet and loosening sexual mores took effort – it meant looking for coded adverts in the backs of newspapers, hanging out at the right spots, knowing what to say and who to say it to, and building a network of like-minded people. The stakes were high: arrests and raids at gay clubs and establishments of 'ill repute' were common, which often came with the public naming of those who were there and arrested. People lost their jobs, families, homes. For some people, however, the risk of losing everything was worth it to be able to pursue their passion in the company of like-minded individuals.

For the better part of modern history, the stigma against people engaging in sexual activities outside the mainstream was justified by a psycho-medical establishment that judged those behaviours pathological, due in no small part to a narrowed understanding of pain as something that is always bad and always to be avoided. In 1886, Austrian psychologist Richard von Krafft-Ebing published his *Psychopathia Sexualis*, which catalogued and categorised certain sexual behaviour as 'perversions' or 'paraphilias' under a medical-psychiatric rubric. Krafft-Ebing coined the words sadism and masochism, in addition to medically describing – in moralising terms – practices such as voyeurism, fetishism and, of course, homosexuality.

Krafft-Ebing's book was a bestseller – probably itself an indication that at least some of the behaviour he labelled deviant was in fact much more common – but it also set a medical and legal context that would persist for decades. Sexual enjoyment of things typically considered painful couldn't be considered anything other than a pathological, psychosexual deviance. This attitude persisted well into the 20th century, with significant negative implications for the people labelled psychosexual deviants. According to a 1998 survey from the National Coalition for Sexual Freedom, founded in 1997 to advance the sexual rights of consenting adults, 36 per cent of S&M practitioners reported being the victims of harassment; even worse, 24 per cent of them had, as a result, lost a job or a contract, 17 per cent lost a promotion, and 3 per cent lost parental custody of a child.

It wasn't until the *DSM-5* was published in 2013 that the American Psychiatric Association officially stopped considering anyone who voluntarily enjoyed pain or humiliation in the pursuit of pleasure mentally ill. The new diagnostic criteria reframes pathology around whether an act or behaviour is causing harm to self or others, though there's certainly some problematic wiggle room in that, it also made it clear that 'the APA's intent to not demand treatment for healthy consenting adult sexual expression.' That it has taken so long to rescue BDSM – along with homosexuality – from the depths of pathology demonstrates psychology and medicine's ongoing struggle to differentiate between pathological behaviour, behaviour that is criminal, and behaviour that much of society simply considers socially taboo.

Negative stereotypes about the people who practise BDSM still persist (and don't even get us started on *50 Shades of Grey*, which is an inaccurate depiction of BDSM and positively terrible depiction of adult relationships). However, things *have* changed, quite possibly owing to the increased mainstreaming of sex-positivity movements, as well as the body of academic research into exactly who practises BDSM. Estimates of BDSM involvement are tricky. Researchers have not agreed upon a precise definition – and, even if they had, many people view the activities differently. This explains the tremendous range of estimates: surveys put engagement in BDSM at between 1.5 per cent and 68.8 per cent of the population, a differential so big as to not be at all useful. We have a slightly clearer picture of how many people might *want* to practise: a 2015 article published in the *Journal of Sexual Medicine* reported on a survey of 1,516 participants who were asked about their sexual fantasies; the results suggest mixing pain and sex is not at all uncommon. When it came to BDSM-related fantasies, 65 per cent of women and 53 per cent of men fantasised about being dominated during sex; 47 per cent of women and 60 per cent of men fantasised about dominating someone else; 52 per cent of women and 46 per cent of men fantasised about being tied up for sexual pleasure; 42 per cent of women and 48 per cent of men fantasised about tying someone else up; 36 per cent of women and 29 per cent of men fantasised about being spanked or whipped for pleasure, while 24 per cent of women and 44 per cent of men fantasised about spanking or whipping someone else. That's a lot of fantasising.

As to what kind of person is fantasising about sex and pain, several studies highlight how 'normal' and healthy BDSM practitioners are. A 2006 study comparing practitioners to non-practising adults found lower levels of depression, anxiety, PTSD, psychological sadism, psychological masochism, borderline pathology and paranoia among practitioners. In the late 2000s, researchers at the University of New South Wales conducted a national telephone survey of over 19,307 Australians between the ages of 16 and 59. In addition to finding that practitioners are no more likely than non-practitioners to have experienced sexual coercion, sexual problems or psychological distress, they wrote, 'Our findings support the idea that BDSM is simply a sexual interest or subculture attractive to a minority, and for most participants not a pathological symptom of past abuse or difficulty with "normal" sex.'

Practising BDSM might also, itself, be healthy or at least attract healthy people. A paper in the *Journal of Sexual Medicine* compared 902 members of the BDSM community to 434 non-members and found that, compared to the general population, the BDSM community was less neurotic, more extroverted, more open to new experiences, more conscientious, less sensitive to rejection and had a higher subjective well-being. And BDSM communities might be more prosocial than other spaces, given how valued consent, communication, care and negotiation are to most of them. One recent study investigating 'rape-supportive attitudes' found that BDSM practitioners scored significantly lower on benevolent sexism, rape myth acceptance and victim blaming, although they scored similar to non-practitioners on measures of hostile sexism, expectation of sexual aggression and acceptance of sexual aggression. The results both demonstrate that practitioners may hold a more supportive view of consent, and challenge the idea that members of the BDSM community are more likely to act out of sexual aggression (they are not, although they may be *just as* likely).

In practice, BDSM groups are deeply committed to keeping public events safe; studies show that most public parties have a 'dungeon monitor' watching scenes to ensure everyone is following the rules.* Illegal drugs are prohibited, and it is a social norm to look out for each other, for example, stopping their own practice to check on anyone yelling a safe word. Negotiations can be as quick as a 'hey, how do you feel about me poking you with this hot iron a couple of times?' to pages-long legal contracts, but they require leaving space for both parties to say yes *and* no. Individuals who don't play by these rules – who do not talk in advance about boundaries, safe words and off-limit content, or who do not arrange for after-care for their scene partner(s) – are often not accepted within the wider community.

What this all means is that finding sexual pleasure in pain is not pathological – and it's not just about sex. Many people who do it are themselves psychologically healthy, do it for psychologically healthy reasons and see it as a path to psychological health.

Tie me up, tie me down

Lillith Deville, 34, has always had a lodestone attraction to physical pain. 'You know that old warning about sticking forks in outlets?' Deville says. 'Yeah, that was me. One of my favourite things to do as a toddler was to unscrew light bulbs and stick coins in the lamp. I just knew I liked the way it made me feel. I don't know why.' Those first experiments with pain soon became a kind of emotional, spiritual, philosophical framework; as a gender-fluid dark arts performer whose act includes trapping his fingers and toes in animals traps and inviting audience members to staple money to his body, his interest in pain also eventually became a paying career and an art form. And a source of pleasure, expressed within the expansive landscape of the fetish community, an inclusive term used to describe anyone who enjoys mixing different 'flavours' of sexual activity into 'vanilla' sex life, including but not limited to BDSM. 'I do get a pleasure out of pain, it is a euphoria that comes from it,' Deville explained. 'It's a little

* People in the BDSM community refer to these instances of negotiated sexual play as 'scenes', so we will too.

drug-like, getting kicks from being an exhibitionist and putting myself through something that is painful.'

But the benefits extend beyond any one scene. Deville says he deals with unwanted physical and emotional pain differently since he's been involved in BDSM. 'When you willingly subject yourself to pain you learn how to process and learn different techniques to deal with it in the real world,' he said. At the same time, having a safe space to express and engage pain is, he says, like therapy: if he finds himself ruminating on something distressing, he can enter into a fetish scene and pour his anxiety into the pain. By the end, he feels 'cleansed'. 'Every time I've come down from rope suspension I've had a whole emotional outpour,' explained Deville. 'It's like a purge almost.' Ultimately, his practice can seem more spiritual than sexual, although it's probably safer to say it's both.

One of Deville's favourite practices is *shibari*, or rope bondage. Though the name, which means 'decorative tie', was only applied to the practice in the 1990s, *shibari* can be traced back to the 13th-century Japanese practice of *hojojutsu*, a form of rope-tying that was both a means of restraint and a form of punishment. Prisoners were detained in ropes tied with complicated knots in configurations that are not only uncomfortable but also tightened with movement, ensuring immobilisation. During the Tokugawa or Edo period, between 1603 and 1868, the practice was combined with eroticism and given the name *kinbaku*.* In modern practice, *shibari* is a skilled, meditative art form and sexual practice that involves binding a partner or oneself with rope tied in intricate knots across precise parts of the body. The process of rigging up safely can take hours, depending on the complexity of the knots; the end result can be aesthetically beautiful but is also, by design, incredibly painful to participants. And that's the point. 'Pain is a challenge to overcome. It's going to be there, then

* This era is characterised as a stable period of growth in the economy, arts and culture, but is also known for its strict social order and hierarchy as well as aggressive exclusion of outsiders. Power and authority were tightly controlled and conferred based on ranking from the Emperor down, with very little social mobility. The introduction of rope-tying into erotic play is almost too perfect. #FreudianFieldDay

what are you going to do with it?' Deville said. 'Are you going to allow it to be something to push your limits?'

Shibari is in the 'bondage' category of BDSM, and any category is limited only by one's imagination and partner's willingness. The foundation of many of these practices is an exchange of power. Sex – or rather, the frisson of sexual contact and arousal – can be part of this exchange but isn't necessarily the focal point. The terms dominant, or top, and submissive, or bottom, are often employed to describe this power exchange, though the terms may not exactly capture the complex, layered dynamics of power and control at play within a scene. As covered in Chapter 5, control is not always a straightforward phenomenon.

As we've talked about, control can reduce the intensity of pain because we know it's coming and can prepare; it stands to reason that giving up a little control, in the context of BDSM practice, can increase the intensity. And this, for some, can increase the pleasure – you don't know *exactly* when the pain is coming, but you really *want it to.* In the context of motivated pursuit, consent and preserved agency, getting the pain we want but not knowing exactly how is like any other highly anticipated and surprising reward. This paradigm – handing control of your body over to another – also offers a different kind of pleasure.

'There's a lot of trust that has to go into basically giving your body to somebody and letting them do what they will with it,' explained Deville. Every suspension, scene and exchange is different, and each person sets their own intent – what they want from the experience – going in. Scene partners help each other reach new levels of emotional connection and intimacy, predicated on a negotiated transfer of control. At the same time, the given consent and option to withdraw it at any time preserves individual agency – their ability to act. It is in this interplay – no control, yet preserved agency – where the boundaries of trust and intimacy can be explored and deepened. The resulting experience is frequently described as euphoric, sensual and intensely vulnerable. And, according to social worker and author Brené Brown, whose lecture on the power of vulnerability is one of the top five most viewed TED talks ever, vulnerability 'is the birthplace of love, belonging, joy, courage, empathy and creativity.'*

* Bet you never thought you'd read a book that linked Brené Brown and BDSM.

It may seem paradoxical that BDSM practices can lead to deeper connection and intimacy, even a sense of freedom. But within the boundaries of a BDSM relationship or even a short scene, prescribed roles and consent offer an opportunity to experience vulnerability and be affirmed in it. Just as a kite on a string soars and loops but doesn't fly away, practitioners can forget themselves in a scene and just *be*, knowing someone else is holding the rope. In describing one of his most memorable encounters with a partner who was also a very close friend, Deville says, 'He knew where I was, mentally, and knew my range and my limits. He was able to push limits and boundaries and it was really one of the few times I was able to completely let myself go.'

Members in the community often describe similar experiences of 'letting go', of feeling both out-of-body but also completely grounded in their body. It's this state of altered consciousness that intrigued Brad Sagarin, social psychologist and head of the Science of BDSM research group at Northern Illinois University. His research team collects data from places like the Arizona Power Exchange, a members-only charitable organisation dedicated to serving the BDSM community, or FetLife.com, the nearly 8.3 million-member social network for kink, in equal collaboration with representatives of the BDSM community.*

Sagarin's lab has conducted several studies exploring the mechanisms behind these kinds of altered-consciousness and meditative states that BDSM practitioners report experiencing. In a 2017 study published in *Psychology of Consciousness: Theory, Research, Practice* the team randomly

* This isn't to imply, however, that the BDSM community is a monolith – there are many practitioners, individuals, groups and sub-groups, all of whom may have different definitions, interests and practices. When we say 'community', we're talking about BDSM-oriented groups that have adopted the 'safe, sane, consensual' (SSC) or 'risk-aware consensual kink' (RACK) principles. Though they are not the same, they are similar, and both seek to offer a structure and framework to the practice. The question of 'safe' and 'risk' are big questions for the kink community, especially given that risk is precisely what some people are interested in experiencing and 'safe', in this context, can be subjective, given the chosen practice. But the important thing is that those engaging in the practice have a clear understanding of what's involved.

assigned 14 experienced BDSM practitioners to either the bottom/submissive position or the top/dominant position, and then asked them to participate in a scene. Based on previous literature and theory, Sagarin and team hypothesised that the altered states described by those in top position – known in the community as 'topspace' – may be similar to 'mental flow state', a state of complete immersion in an activity first described by psychologist Mihaly Csikszentmihalyi. The bottoms' self-reported 'subspace', they suggested, may in turn result from transient hypofrontality, which is a decrease in neural activity related to executive functioning – basically, the 'getting out of my head' experience.

Their findings revealed that, actually, *both* tops and bottoms scored high on the flow state scale following the scene and, as predicted, bottoms demonstrated characteristics of transient hypofrontality. Self-reported data revealed that negative affect went down for both positions from before to after, but that positive affect only significantly increased for tops. These findings suggest that BDSM can actually be a pathway to being mindful. 'BDSM, because of the intense sensations and potentially because of the restriction of movement, may have the ability to put someone in the here and now in a way that they may find more difficult to achieve through other means,' Sagarin told *TIME* magazine in 2016.

But there's more. This state of mindfulness is, in part, supported by the practice's power dynamic. The sensation of 'letting go' may work to bring scene partners closer together: Sagarin's studies show that participants' assessment of 'self-other' overlap increases from before to after sharing a scene. Self-other overlap is essentially a measure of integration, or the degree to which a person's sense of 'self' is independent verses merged with an 'other'. These findings suggest that scene partners 'lost themselves' in each other, creating a powerful feeling of closeness.

This last finding, that scene partners feel more immersed in each other, suggests a deepening of empathy. And empathy, according to those in the community, is like BDSM's secret sauce, for both bottoms and tops. The 2015 study in the *Canadian Journal of Human Sexuality* we mentioned earlier found that, 'Dominants and submissives did not differ on empathy, honesty-humility, conscientiousness, openness to experience, altruism, or agreeableness.' In fact, empathy is one of the most important traits scene partners look for – according to the article, community members wanted dominants who were 'empathic and nurturing, desiring and able to take control, and attentive and responsible'.

Broadly, empathy is the ability to put yourself in someone else's shoes in an attempt to experience what you believe they are experiencing. Empathy can be cognitive, as in trying to work out what someone else is thinking, or it can be somatic, experiencing someone else's physical sensations – for example, feeling pain in your tooth after seeing a friend crack theirs on a piece of hard toffee. Just like everything else, an experience of empathy is constructed by forming an internal model based on predictions. To get an idea of what we think they are feeling, we simulate the predicted experience within ourselves; evidence from fMRI supports this idea, showing similar neural network activity when we experience something and when we watch someone else experience the same thing. However, evidence also shows that while there are overlapping networks, there is also distinct activity, for example, in the interoceptive network. After all, we may feel the pain of our friend's broken tooth, but we're also processing all the activity in our own body and, of course, we haven't actually broken our own tooth.

Our prediction of what someone else is feeling is never going to be an exact match, it's just our best guess. But it's usually good enough, and this empathetic force is what holds partners together through a scene. Negotiated consent provides the foundation of trust to enter into a scene, but it is maintained and deepened through empathetic communication. As Deville told us, it's knowing how to read changes in your partner's breathing, their eye contact, vocalisations, the meaning of every wince and utterance. 'One of the subtle things I enjoy about being a fetishist is reading the submissive, reading the bottom, and just trying to interpret through their body language, their facial expressions, what their sounds are like, where they're at,' explained Deville. But, he added, 'Can't be a good top until you're a good bottom.'

Through co-creating a safe, supportive, consensual space, BDSM players are able to explore and push the boundaries of what it feels like to trust unconditionally and to be completely vulnerable. Here, the pleasure from pain is dependent upon having not only a willing partner but a partner whose needs are also being met. This doesn't always mean enjoying the pain precisely. In fact, the pain might be tremendously unpleasant and yet the entire experience still enjoyable: the joy may come in knowing how much participation means to a partner, or the pain may be a symbol of commitment, dedication, trust and love. For some, the delight is in knowing that their partner is dependent upon them, their pleasure is in their control and they could take it away at any second. Ultimately, there are countless pathways to pleasure, and discovering them is part of the excitement.

Let it go, let it go

There is, it seems, a lot to be gained by practising safe, sane and consensual BDSM. Don't just take our word for it – the 2015 study in the *Canadian Journal of Human Sexuality* highlights the diversity of rewards members report from their involvement: 'The general benefits of BDSM mentioned by the participants for both roles were pleasure from pleasuring others, physical pleasure and arousal, fun, variety, and going beyond vanilla, personal growth, improved romantic relationships, community, psychological release, freedom from day-to-day roles, and being yourself.'

Brad Sagarin's research has led him to suggest that those into BDSM have unlocked a wider emotional palette and learned just how much more they are capable of feeling. They seem to have tapped into using pain as a way to find pleasure and enjoyment, to find peace and that millennial buzzword, *mindfulness*, in part through negotiating power dynamics with others, building intimacy and vulnerability in those spaces. They're able to do that because they've opened themselves up to the possibilities of pain not being a bad thing, and in doing so reap the neurochemical and emotional rewards. But more than that, they've learned how to make their own meaning from their experiences. They've set aside ideas about what is supposed to feel good or bad and instead, just *felt*.

This doesn't mean everyone has to get on board with being strapped to a board. But how many experiences have you denied yourself because the little voice in your head or 'society' told you it was weird and that you shouldn't like it?

It's time to let that go. Our bodies are capable of so much more than the narrative of pain always being bad, always being scary, always being something to avoid allows. Just ask the women who've orgasmed during what's supposed to be the worst pain of their lives. What we have to stop doing is allowing the experiences and opinions of others – even well-intentioned psychologists with questions about why we like hot chillies and whisky – dictate the meanings we make. Instead of attempting to suppress or deny what our body is capable of feeling, why not – with these new-found skills in interoceptive awareness and agency and always within the context of consent – dive in and see what feels good? Who knows? You might just like it.

CHAPTER EIGHT

I Feel Your Pain

In January 2006, Scott Bell earned a place in the *Guinness Book of World Records* for the longest distance walked over hot coals. It was 250 feet. And he was in agony.

Fire-walking isn't usually painful – it's one of the great secrets of the experience, that our fear of the pain is actually greater than the pain itself. In fire-walking, hardwood is burned down to its coals and though the coals themselves are very, very hot – more than 500°C, in fact – they conduct heat slowly, especially compared to, say, metal. This isn't to say that the risk of tissue damage is non-existent, but if the coals are prepared correctly and the walker doesn't stop to take a selfie, they're not likely to get burned.

But 250 feet is a long way, nearly the length of a football pitch. And Scott did get burned. He set out twice to walk the distance, each time stepping off; it was too hot. 'I got some pretty big blisters and I was in quite a lot of pain,' he recalled. Self-doubt was winning. After he'd stepped off the second time, his wife approached him. 'One more time,' she said. 'Just try it one more time.'

'The pain just went at that stage,' he said. The third time was the charm: he did it. And then nearly a year later, in November, Bell did it again – this walking 328 feet, smashing his own world record. (His record, incidentally, was broken the following year by a man in Canada who walked 593 feet. As of September 2019, the world-record holder is Hungarian Csaba Kerekes, who walked 200 metres, or 656.61 feet, over hot coals.)

Bell – along with his wife and daughter – now runs a company in Manchester that facilitates fire-walking experiences for individuals and groups, often either charities looking for a unique fund-raising event or companies looking for a unique team-building exercises. The practice of fire-walking stretches back to at least 1200 BC, and is found in cultures around the globe; in historical texts, fire-walking is cast as a healing exercise, a public test of strength or a feat of mental clarity.* Not unlike now. Since the 1980s, corporations that felt like ropes courses and trust falls simply weren't powerful enough have used fire-walking to 'unleash' their employees' potential; they were inspired by motivational guru Tolly Burkan who posited fire-walking as a kind of metaphor for overcoming self-imposed limitations. Now fire-walking is being joined by an increasing number of 'extreme' team-building exercises, including glass-walking, board-breaking and, even more recently, Lego-walking.

Notably, most of these are exercises that look painful and can be painful but typically aren't associated with injury.† But that's part of what makes them work so well, on a lot of levels. The point is that we really believe it might hurt – when we can see the sharp, glittering points of glass, we know the board is hard, we can feel the waves of heat rolling off the hot coals, we are buoyed by the mixture of fear, anticipation and excitement. And just the fact that it might not cause damage doesn't mean our predictive brain won't start firing off exciting endogenous opioids in anticipation of the pain. So when that pain doesn't happen, and we're also pumped up by the thrill of the experience, then we feel even more amazing, we feel superhuman. At the same time, any tissue damage that may occur – blisters, for example – is often reframed as something positive; one fire-walker we spoke to called blisters 'kisses'. Equally, with something facilitated and controlled like fire-walking, individuals also retain a degree of agency over the experience – they choose when and if they start on the path.

---

* And a way to dodge the draft: Roman historian Pliny the Elder, writing in the first century, recounted, 'At the yearly sacrifice to Apollo, performed on Mount Soracte, walk over a charred pile of logs without being scorched and who consequently, under a perpetual decree of the Senate, enjoy exemption from military service and all other burdens.'

† Bell regularly does fire-walking, glass-walking and Lego-walking as part of his company's demonstrations: 'Out of the three that I do on a regular basis, it's before I step on the Lego that I think, "Oh, this is going to be a bit uncomfortable,"' he told Linda for a 2018 *Smithsonian* magazine article. Why? Legos are made from ABS plastic, an extremely hard and durable polymer with absolutely no give at all. They have loads of sharp edges but, unlike glass, seem to resist flattening and shifting. And plus, there's always that one random one – the Luke Skywalker mini-figure, the piece of Batman's cave – that just won't go down.

But that's not all! Fire-walking and experiences like it are not just private challenges, they're shared ones too – part of the draw and the reward is in doing them with other people. These ingredients – the agency, the risk, the excitement, the act of voluntarily gathering with a collective focus around a shared activity of sacrifice and celebration – make for an experience that is uniquely bonding. 'It's not *that* risky … and everyone enjoys watching people be a bit uncomfortable,' said Bell, adding that participants seem to especially enjoy it knowing that they're about to do it themselves: 'It's this "We're all in it together."'

Friendship and connection is built on the bedrock of shared experiences; we know this intuitively. There are lots of behaviours that reinforce group cohesion and what's called 'identity fusion', a sort of merging of the self into the group whole: marching in step, dancing, playing music, singing and chanting in unison have all been shown to support cooperation among members. The emotional intensity that can occur in groups has long been a topic of interest for sociologists and anthropologists. In 1915, French sociologist Émile Durkheim, one of the grandfathers of sociology, described this closeness and immersion in the group as 'collective effervescence', which he describes in his text *The Elementary Forms of Religious Life*: 'The very fact of congregating is an exceptionally powerful stimulant. Once the individuals are gathered together, a sort of electricity is generated from their closeness and that quickly launches them to an extraordinary height of exaltation. Every emotion expressed resonates without interference in consciousness that are wide open to external impressions, each echoing the others.'

Collective effervescence can be understood as an emotional concept, characterised by intense feelings of belongingness and immersion of the self into the larger group. Inventive research in just the past 10 years reveals how these experiences bring people together, how they promote prosocial behaviour, group cohesion and solidarity, and assimilation. For example, emotional and even physiological synchronicity have also been observed at a number of major communal life events. One novel but very small study found that during a wedding, oxytocin levels rose in the bride and groom, as would be expected, but also the mother of the bride, father and brother of the groom, all the relatives tested. And, as history and recent research support, the more intense the ritual, the more powerfully it binds groups together. What makes fire-walking an effective team-building exercise is that it is a shared *intense* sensation – basically, that it involves pain.

Going to extremes

Extreme rituals, according to social science researchers, 'are characterised by physically demanding tasks performed within a social context, usually with fellow practitioners or interested observers.' These kinds of rituals, which frequently involve pain, have been observed in most if not all cultures across history and up to the present day. They serve different explicit purposes depending on context. For example, bloodletting rituals found in Mayan society between 250 and 900 AD were a required rite for Mayan kings when they took the throne; they would, in some cases, pierce their tongues or penises or cut their skin, then scatter or burn the collected blood. (They also used a stunning variety of tools to accomplish this task, among them stingray spines, obsidian and flint blades, bone awls, cord and bark paper – we've always done weird shit to our body.) These practices weren't restricted only to the assumption of power: Mayan kings and noblemen would be expected to make themselves bleed during other significant occasions, including agricultural ceremonies, communicating with ancestors, procreation and, of course, before, during or after warfare.

Blood rituals were costly, not only for the person experiencing them but also for the survival of a civilisation: staying alive was hard work and infection was no joke, so to intentionally cut up your body, especially your genitals, was risky. And that was the point. Jessica Munson and colleagues from the University of California analysed 2,480 Mayan hieroglyphic texts and found that bloodletting rituals among Maya nobles were primarily recorded within the context of warfare and conflict. During times of conflict, 'buy-in' from the group was essential; bloodletting rituals functioned as 'credibility enhancing displays', demonstrating the depth of leaders' commitment to the group, while at the same time promoting engagement from the group – it's difficult to feel all right about sitting at home while your leader is publicly poking holes in his penis and bleeding for you and your people. Plus, these grim rituals served as a demonstration of power to outsiders: if we're willing to do this to ourselves, what do you think we'll do to you?

The inclination to seal loyalty pacts or to demonstrate commitment in our own blood is deep in the structure of many societies because blood, pain, sacrifice *means* something. If you're seeing echoes of, for example, gang initiation ceremonies or hazing on sports teams and fraternities in modern cultures around the world – being beaten, verbally abused, pelted with lit fireworks, made to drink alcohol or take drugs, stabbed, raped or forced to run a gauntlet – well, you're right. However, this type of shared but coerced experience around pain or the threat of pain leaves real long-term physical and psychological scars, and of course it is criminal. Just as we highlighted in the previous chapter in our discussion of BDSM, in order to harness the real prosocial benefits of shared intense experience, enthusiastic consent is crucial.

Though leaders in the modern world are no longer expected to bleed for their people, and we don't routinely sacrifice humans or animals (well, now we pair the animals with some stuffing and gravy and call it a holiday), we're still finding ways that shared suffering brings us together.

One tradition that has stood the test of time, and has gained the attention of social scientists, is the Hindu festival of Thaipusam. In it, devotees carry kavadi, or 'burdens', as an offering and show of devotion to Lord Murugan, who represents virtue, youth and power. Some carry a kavadi to symbolise the burdens for which they are asking assistance, others carry them as a promise of suffering in return for a gift, such as the birth of a child, from the god. The kavadi may be small, such as a heavy container of milk, but for others, the kavadi consists of piercing their cheeks, tongues, chests and backs with skewers called 'vels', after Lord Murugan's ceremonial sword. Those with the heaviest burdens may carry an elaborately decorated altar, attached to their body with 108 vels pierced into their skin.

Their actions appear excruciating, yet – in accordance with the Hindu practice of detachment, which separates our bodies from the earthly world – the devotees report no pain. In the Hindu tradition, pain is not random nor is it punishment; rather, it is the soul's unfolding of karma, a consequence of previous behaviours and a condition to be accepted completely as natural. Pain is given by the gods and therefore a gift that can be used to further spiritual growth – devotees report being flooded with spiritual enlightenment and euphoria.

Thaipusam, along with other modern-day extreme rituals including fire-walking, has been studied by only a handful of social scientists, among them Ronald Fischer from Victoria University Wellington and Dimitris Xygalatas from Aarhus University. Independently and together, they have published research gathered across several years and a variety of extreme-ritual settings; their work offers insight into how extreme rituals – and, more specifically, pain within those rituals – serve to promote social cohesion. Their research argues that extreme rituals have socially evolved to promote in-group cohesion and strengthen social networks. In a 2014 paper, Fischer and Xygalatas explained, 'We argue that pain and suffering are specific, biologically rooted emotional experiences that are particularly powerful in binding groups together.'

In one study, published in a 2013 paper, Xygalatas collected self-reported data from 86 men who participated in a Thaipusam festival in Mauritius, a multi-ethnic island nation in the Indian Ocean: 19 men who engaged in 'a high-ordeal ritual involving body piercing with multiple needles and skewers, carrying heavy bamboo structures, and dragging carts attached by hooks to the skin for over 4 hours before climbing a mountain barefooted to reach the temple of Murugan'; 32 high-ordeal observers who walked alongside the performers; and a low-ordeal group of 35 who had participated in collective prayer three days prior to the ritual. Following their ritual engagement, participants were asked a series of questions designed to measure prosociality, or the degree to which individuals felt empathy with and responsibility for members of their group. Xygalatas found that those in the high-ordeal groups – those who performed and those who observed them – endorsed the more inclusive, national Mauritian identity compared to the low-ordeal group, which was more likely to identify with the more parochial ethnic-religious Hindu identity. Pain was also significantly related to identity; higher pain ratings were positively correlated to embracing more inclusive identity.

The data also showed that those in the high-ordeal performance and high-ordeal observer groups donated significantly more to charity than the low-ordeal group, with those in the performance group donating the most. Xygalatas found that pain was significantly correlated with donation amount – the more reported pain, the higher the donation, even after controlling for age, religiosity and temple attendance. This tallies with other research, albeit in less dramatic settings, demonstrating that adverse or nociceptive experiences can make events more meaningful for individuals participating in them. In 2013, researchers at the University of Warwick in the UK and Princeton University in the US found people donated more to charity when they thought they'd have to suffer to do it – running five miles, for example, or plunging their hands into ice-cold water for 60 seconds. The findings held up across five different experiments and researchers concluded that a painful or challenging experience seemed to allocate more meaning to the act, making participants more engaged in it. They called this 'the martyrdom effect' and noted, '[W]illingness to contribute to a charitable or collective cause increases when the contribution process is expected to be painful and effortful rather than easy and enjoyable.'

Research also seems to demonstrate that the positive effects stretch beyond the individual experiencing the pain, through a kind of emotional, biological synchronicity. Xygalatas led a research team to a small village in northern Spain to the annual Paso del Fuego, a fire-walking ritual that has taken place in the town every summer solstice since time immemorial. In the Paso del Fuego, more than two tons of oak wood is burned for four hours to produce a carpet of fiercely glowing red embers, reaching 677°C. When the coals are ready, the walkers parade – with music and clapping – into the village's open-air amphitheatre full of spectators. They dance around the coals before settling down to wait their turn to walk. Each walker is heralded with the braying of a trumpet and most carry a relative or loved one on their backs across the 7-metre bed of coals, spending an average of five seconds. Xygalatas's team collected heart-rate data from 12 fire-walkers, nine friends or relatives of the walkers – not those who rode on the walkers' backs – and 17 unrelated spectators visiting the village. The results of the analysis showed that heart rates of friends or relatives, though not unrelated spectators, synchronised with their participating fire-walker.

Undoubtedly, there are neurochemical underpinnings for why pain can both make these experiences more meaningful and make us disposed to feel good about the people who have experienced them with us. During these high-arousal states our bodies are pumping out all kinds of neurotransmitters and hormones including dopamine, endorphins, oxytocin, vasopressin and serotonin. Some of these, such as oxytocin, have strong prosocial bonding effects, making us feel closer to those we're with (although they also can promote defensive behaviour towards those perceived as threatening). At the same time, some of these neurochemicals – specifically dopamine, serotonin and norepinephrine – are also implicated in how we make memories, particularly strong memories. The upshot is that not only do we feel closer to the people we've experienced something intense with but we also remember it all better. Xygalatas's work illustrates the possibilities for empathetic identification and supports the theory of collective effervescence thanks in part to our predictive processing and ability to create a mental model of what our loved ones are experiencing. Everyone shared strong emotional reactions in synchronicity, just to different extents.

The benefits of extreme ritual are not only observed among religious groups or only within the context of a long-held, sacred tradition. 'The Dance of Souls' is a hook-pull ritual and celebration started in 2004 and held at the Southwest Leather Conference in Phoenix, Arizona; during the ritual, individuals pierce their bodies and then attach weights or ropes to create tension against the temporary piercings. In 2017, members of the Science of BDSM research team under the supervision of Brad Sagarin, who we introduced in the previous chapter, completed a comprehensive, multi-disciplinary study of the Dance of Souls. As the researchers expected, negative affect and stress ratings went down significantly from before to during and after the ritual. One participant told the team, 'The experience was quite calming. The degree of pain I allowed myself to have physically allowed my hyper-analytic brain to focus/center on the present. With the exception of knitting and crochet, pain this intense is the only way I can truly fall into this state of peace/have a quiet mind.'

But this 'calm' was also married to a sense of belonging: participants' sense of closeness – measured via the 'Inclusion of the Other in the

Self' scale, the most frequently used measure of how close a respondent feels to another person or group – also increased from before to during and after the dance. One individual told researchers that they chose to participate in the Dance 'for the ecstatic experience, which is always spiritually cleansing and renewing for me, and the chance to share this with my chosen family/tribe.' To feel connected is a joy; to feel not only welcomed but also enmeshed in a collective where you can see parts of yourself in others and feel the 'energy' of the group inside the self can be euphoric. For a group that is often stigmatised, finding each other and celebrating it through ritual can feel empowering.

Pain as social glue – and social contagion

What fire-walker Scott Bell facilitates for corporate IT teams from Devon and charity challenges in Macclesfield is not as intense as ritual hook pulls and facial piercings, and that's probably for the best – as Xygalatas noted in a 2019 article, extreme rituals 'pose significant risks such as injury, trauma, or infection.' HR definitely wouldn't approve. But a little discomfort can go a long way.

Brock Bastian, the University of Melbourne psychologist we met in Chapter 6, also investigates the role of pain in social groups and settings, although in somewhat less extreme or exotic circumstances than Xygalatas, Fischer and the Science of BDSM team. Bastian suggests that pain, in some contexts, acts as 'social glue'. This is a phrase that began popping up in the last two decades to describe a force that bonds individuals and groups together in strong ways; lots of things can be social glue – memory, imitation, moral values, shared beliefs are all examples, but so are some of the technological artefacts of life, such as social media, cheap phone plans and memes.

In one of his research experiments, Bastian asked groups of strangers to engage in a few nociceptive and aversive tasks: plunge their hands in ice water to retrieve some balls, eat hot chilli peppers, and hold a series of leg squats, while a control group engaged in similar but not nociceptive tasks, such as fishing balls out of room-temperature water. The groups that shared nociceptive experiences demonstrated not only increased perceived bonding – they rated their closeness higher – but also increased cooperation during an economic game involving group pay-offs for individual decisions. The control group, on the other hand, didn't. 'Our findings shed light on the social effects of pain, demonstrating that shared pain may be an important trigger for group formation,' he and his team concluded.

In November 2018, Bastian and fellow researchers published a second study that examined how this phenomenon affects creativity in a team setting. In experiments again using hot chillies and leg squats, they demonstrated that 'sharing an adverse (vs. non-adverse) experience leads to increased supportive interactions between team members and this in turn boosts creativity within a [newly created] team.' The researchers added, 'There is a current trend toward making office environments fun and comfortable, and team building exercises revolve around sharing in positive rewards. There are clearly benefits to these approaches, but our work suggests that aversive experiences may be an important, yet frequently overlooked, avenue for achieving similar ends.'

In conversation, Bastian cautioned that the effects he saw didn't imply that pain was *better* than pleasant intense experiences at promoting bonding. 'I don't think any of our effects show that it's explicitly anything to do with pain – sharing pleasant experiences with people can also be incredibly bonding,' he said. 'It's not that pain can explain these things better. It's more just shining a different light on the effects of pain.' And it's also important to acknowledge that there is evidence pain doesn't work better than other rituals to knit groups together. For example, a study published in 2018 in the *European Journal of Social Psychology* examined the social cohesion effects of a painful Brazilian jiu-jitsu promotional ritual, drawing on the self-reports of 605 participants. They found no difference in identity fusion levels between those who had run a painful belt-whipping gauntlet as part of their promotion rituals and those who had not. They also found no positive correlations between painful experience and inter-group relationship bonds or pro-group sacrifice.

However, there is some pretty compelling evidence that even if pain isn't the *only* way to achieve intense emotional bonds, it's one of the quicker ones. Margee, in designing thrill-seeking attractions, found that even the suggestion of intense pain was enough to bond some people together – in one case, for life. In 2015, Margee designed and ran an experience she called 'Sewn Into You' at an immersive masquerade dance party held in a defunct penitentiary in Philadelphia. Couples were told they were about to be 'sewn together in an eternal bond'. After they signed a waiver, proclaiming themselves in good health and not likely to sue, they were blindfolded and asked to hold hands. The lead actor in the scene would then 'prepare' the skin on their hands with alcohol swabs, before making small 'incisions' – light pokes – with a needle. The needle was electrified using a modern-day violet ray, a device that allows electric current to flow through the body, and the sensation realistically imitated the feeling of being cut. The actor then applied a plaster loaded with a cotton ball soaked in red food colouring (an additional surprise when they remove their blindfolds) over the incisions. Then finally, the *pièce de résistance*: the 'sewing'. The electric needle passes under the plaster and the metallic thread is pulled through, eliciting the sensation of resistance and burn, or what would be expected if their skin actually were being sewn together. Thanks to the magic of the electrical current passing through them, each partner could feel when the other was being cut or sewn.

The 'ceremony' closed with a reminder that they were now eternally bonded, a piece of each other forever with the other. But contrary to what you might expect, guests rarely screamed. In fact, queues for the experience were out the door and many couples told Margee later that it was one of the most powerful bonding experiences they had ever had together. One couple even told Margee that the sewn-together ritual was the reason they asked her to be the officiant at their wedding.*

A big part of what made Sewn Into You such a success and a delight was the power of predictive processing; though none of the participants had ever been sewn together before, the cues of alcohol swab, needle, even the waiver, worked together to prime them for the sensation and helped them create the reality. It wasn't the 'thread' that bonded them so much as the pain they experienced together.

---

* Margee is available for weddings, immersive Valentine's Day parties and extremely intense first dates. And team-building exercises!

So we've amply demonstrated that the shared experience of pain can bring people together, and this can have some healthful benefits for communities and individuals. But if pain can be shared, it's worth asking whether pain could also be contagious. We use the word contagious here figuratively; as we've said, our emotions are constructed in time and place based on sensory input, experience and context, and now we're adding another input: other people's emotions.

We know that certain behavioural and psychiatric disorders can be contagious – figuratively speaking. In the last few years, for example, mass psychogenic illness was blamed for the strange behaviour of dozens of teenage girls in upstate New York seized by uncontrollable tics and Tourette's-like outbursts. Outbreaks of terrifying, physically painful yet physically impossible conditions are also characterised by contagion: epidemics of *koro*, also known as genital retraction syndrome, have been observed in multiple South East Asian countries in the 20th and 21st centuries. In *koro*, sufferers believe that their penises, nipples or vulvas are 'retracting' or shrinking into their bodies; as news of the cases spread, more and more people become 'infected'.

Koro and mass psychogenic illness are fairly spectacular examples of how we communicate states of being, but we have all experienced the ways emotions are contagious. Have you ever been in a fine mood until your partner comes home in a terrible mood? Have you ever found yourself feeling a bit lighter, a bit more joyous after sitting in the happy, roaring crowd at a sport ball game?* Found yourself crying because someone in a film is crying? Reams of evidence suggests that emotions are very contagious, and that we don't even have to be in the same room to 'catch' them – just the same social network will do. One 2014 study of nearly 700,000 Facebook users found that when positive expressions were reduced in their news feeds, people produced fewer positive posts and more negative ones. When negative expressions were reduced, the opposite occurred. 'These results indicate that emotions expressed by others on Facebook influence our own emotions, constituting experimental

* We are not big sport fans.

evidence for massive-scale contagion via social networks,' the researchers wrote.*

Emotional contagion makes a lot of sense from an evolutionary psychology standpoint – our survival depended on being able to function and cooperate in groups, and rapid emotional communication facilitates that. Rage, anxiety and fear are among the most quickly communicated emotions, also for good reason: if another tribe's raiding party is spotted on the horizon, those are precisely the emotions that might motivate your group to take action. It also underpins some of the biological and emotional synchronicity observed in prosocial rituals, in which people are feeling each other's joy, pain and catharsis.

Pain, as an emotion, could quite possibly be communicated in the same way. In the 1980s, for example, Australia saw an epidemic of wrist and hand pain, attributed largely to repetitive stress injury. The problem was that there were no concurrent changes in behaviours to explain the rising numbers of people suffering. Some psychologists suggested that, rather than there being a mechanical origin, the pain the Australians were suffering was somehow communicated from one person to another. Similarly, one 2008 study from Germany tracking back pain in people in West Germany and East Germany before and after reunification found that rates of back pain in East Germany rose to meet the same levels as in West Germany. The culprit, researchers suggested, wasn't so much change in lifestyle as it was who East Germans were talking to: back pain could be communicable, if communicable could 'refer to something being transmitted by sharing or exchanging information.'

---

* Though the research is useful in demonstrating how Facebook and social media can and does manipulate users, this study violated a number of ethical guidelines; the journal it was published in, the *Proceedings of the National Academy of Sciences,* published an editorial expressing their concern regarding the lack of ethical considerations. Facebook users had no knowledge or opportunity to opt out of this study, which intentionally manipulated their news feed to investigate how it changed their behaviour. Authors got around the Institutional Review Board because the experiment was technically conducted internally by Facebook. We've learned not to expect much in the way of ethical consideration from Facebook, but for those who are trained in human subject research, as these authors were, and who choose to ignore it – well, that's transgressing some serious professional and civil boundaries.

All of this underscores the fact that pain is shared – for good or for ill. And sometimes, or often, it's shared in *causing* another's pain.

The last execution

On 13 August 1936, every hotel in Owensboro, Kentucky was booked. People who couldn't get a room camped in the fields around the town, pitching tents where they could. In one day, more than 20,000 people poured into the small town, doubling its population overnight. By dawn the next day, the crowds had swelled even more, as more people drove in. As the day wore on, vendors peddled hot dogs and cold bottles of lemonade to the sweating throng as they craned their necks to see Rainey Bethea's snap.

Bethea, a Black man in his mid-twenties, had confessed to raping and murdering a 70-year-old white widow, Lischia Edwards. Kentucky was the last state in America to permit public executions, and these were reserved for only the most heinous crimes – those that were a threat to the very fabric of society and required public enactment of punishment. This is the other way pain holds society together, an effect that is just as powerful as it is distasteful: when the pain is inflicted on a third, mutually chosen party.

In this case, it wasn't only the lurid nature of Bethea's crime that drew the crowds. In the weeks before the hanging, the press got hold of the fact that it would be a woman, Sheriff Florence Thompson, pulling the lever to release the trapdoor under him. Thompson, a mother of four, had stepped into the role just four months earlier when her husband, Sheriff Everett Thompson, died of pneumonia. Papers as far away as Paris reported on the 'Lady Sheriff', the 'hangman in skirts', expected to 'send Bethea into eternity'.

Though she'd told the press that she didn't want to 'inflict an unpleasant job – really my own hard task – upon someone else', Thompson ultimately decided to accept the offer of a man who'd volunteered for the duty: former Louisville police officer Arthur Hash, who wrote to Thompson that he wouldn't want his own mother to be placed in such a position. He was assisted by Phil Hanna, an Indiana cattle-breeder and 'consulting expert executioner' with a passion for public hangings. Hanna travelled with his own portable gallows, and his job, as he saw it, was to set up the rigging, tie the noose and assure the convicted that death would be quick and painless. Bethea's execution was at least Hanna's 70th. Hash had asked for anonymity, then showed up on the day wearing a white linen suit and a panama hat, an ensemble that later earned him the description 'best-looking cop' in Louisville by local press; he also seemed drunk.

As the spectators pushed closer, Bethea mounted the steps to a platform some 12 feet over their heads. What actually happened next isn't clear. Some reports claimed the crowds booed the priest as he performed the last rites, cheered as Bethea's body plummeted through the trapdoor, and swarmed the gallows for 'souvenirs', tearing off his black hood. Other reports claimed a hush had fallen over the crowd when Bethea died, and witnesses remember simply standing and watching. An editorial in the Owensboro paper the next day declared it wasn't the crowd but the bloodthirsty press 'from cities where nothing is cared about the horrible crime Bethea committed' who wanted souvenirs 'to show the half-civilized readers of their yellow sheets.' However it happened, the media circus surrounding Bethea's hanging was deeply embarrassing to the Kentucky legislature; the next two death sentences were carried out in private, before the State Assembly was able to meet in 1938 to introduce legislation outlawing public executions. Though he signed the bill into law, Kentucky Governor Albert B. 'Happy' Chandler did it with regret, declaring, 'Our streets are no longer safe.'

It was psychologist Delroy Paulhus at the University of British Columbia who told us about the images from Bethea's execution; Linda interviewed him for an article she wrote for the *Boston Globe* about evil. 'It's almost like the Beatles' last concert – there's too many people there to actually make it practical to have a public execution,' Paulhus told us. The perception of justice being served in this case enabled spectators to fulfil what might be termed sadistic urges: 'I think there are a number of situations that tend to provoke or, shall I say, allow us to enjoy our dark side … Any kind of punishment, to a large extent, seems to make us feel better to the degree that someone is suffering and it's permitted because the person deserved it.'

Culturally, socially, we are primed to understand pain as punishment, whether inflicted on us by others, self-inflicted or, as in the case of

Rainey Bethea, when we inflict it on a perceived transgressor. As we discussed in Chapter 4, corporal punishment is a damaging but also largely ineffective way of disciplining children; it probably, contrary to what Governor Chandler worried about, doesn't do much to curb adults either. And yet, hurting other people out of a sense of justice has deep roots. Capital punishment is legal in 30 US states; caning, using a thin length of rattan to whip individuals convicted of a crime, is a fairly common punishment in Singapore, Malaysia and Brunei for everything from petty vandalism to armed robbery to sexual assault. We may look back at justice systems that relied on trial by ordeal – as in the infliction of painful torture to determine guilt or innocence – as barbaric, but even now, major 'civilised' world powers use torture under the guise of 'enhanced interrogation techniques'. We know that hurting people doesn't curb behaviour, capital punishment doesn't keep other people from committing crimes, and mounds of scientific and historical evidence suggests inflicting pain is a terrible way to get to the truth. But we still seem to believe pain can be, even should be, an arbiter of justice.

Though it is not a trait most of us are eager to admit to, studies show we will go to great lengths to feel we have delivered justice to those we believe behaved unfairly. Say you're given £10 and told you can give as much as you'd like to another person. If they accept your offer, you both can keep the money. If they refuse your offer, you both lose the money. The other person has no motivation to refuse the offer, other than to punish you for a perceived injustice. In an emotionless, strictly rational world, that person would accept any offer because something is better than nothing. But that is not what happens. Studies routinely demonstrate that those who are offered £4 or less will refuse the amount as a punishment to the giver for not being fair. And we won't just deny money – other studies show that participants will administer stronger electric shocks to punish someone for a minor perceived injustice. (The lesson here? Always split it 50/50, you selfish bastard.)

We internalise social norms early, including ideas about what is appropriate to do and feel when those norms have been breached. One study from 2013 found that children as young as four years old experienced *schadenfreude*, the pleasure at another's person's misfortune, and that this was especially acute when they believed the individual was somehow at fault for this misfortune. Another study from 2018 demonstrated that six-year-old children would rather spend money on watching an antisocial puppet be punished than spend it on stickers.

Why is this? The answer requires understanding how pain is being used in these contexts, both at the individual and at a societal level, and what it is achieving. In 1893, Émile Durkheim, the man who coined the term 'collective effervescence', famously wrote that it is the deviants who hold society together. What he meant is that when members of society transgress a shared social value, this act actually serves to reinvigorate support and protection of that value, as well as instil an appreciation for and renewed commitment to the community as a collective. Through demonstrably punishing transgressions, our social norms, our values, morals, all the threads that hold our identity, our worldview and our society together are strengthened, reinforced and inculcated.

First, it's worth just taking a moment to point out the obvious: we get mad when people do harm. Whether it's instrumental harm (as in hurting creatures great and small, people, places and things) or symbolic harm (as in violating the social contracts and values between individuals and institutions in any given society), it pisses us off. Anger is a highly motivating emotion, and historically this anger motivates us as individuals and collectively as a community to fix what was harmed. One path is through restoration, such as repainting a profanity-riddled work of public art, returning stolen items, rebuilding in the wake of an attack. Another is removal, keeping people away from those they mean to harm.

But so many of the harms we experience are not so easily repaired: the loss of trust following a betrayal, the trauma from an assault, and nothing can replace a life that was taken. When the significance of the value that was transgressed must be reconfirmed, and it cannot or does not happen through restoration or even removal, then we seek it through retribution or punishment – the proverbial eye for an eye. As with so many things in life, power is at the centre of this exchange: the person who has transgressed has positioned themselves above the written and unwritten laws. Retribution aims to reverse this, knocking the transgressor down the social and moral hierarchy through humiliation, deprivation and often pain. When this takes place in front and with the support of the impacted community, the shared commitment dedicated to punishing the transgressor re-establishes the significance of the value transgressed, and reseats power where it is rightfully deserved. This reaffirmation of social mores and codes of conduct is rewarding for the community, increasing solidarity and unity.

In addition to the reaffirmation of social values, the goal is often to elicit remorse in the transgressor. The question that has interested social scientists, philosophers and theologians for centuries is why it is important to us that the transgressor feels bad about what they have done. Because, apparently it is. Remorse is critical for mending conflict, forgiveness, gaining trust, social acceptance and moral standing; we learn this from early on, when our parents demand we *apologise* for whatever we've done wrong, when the church asks us to repent and seek forgiveness. Academics define 'remorsefulness' as an acceptance of wrongdoing and the violation of shared rules or values in a manner that communicates humility and self-degradation, usually expressed as an apology. Studies reveal that when transgressors show they understand the reason for their punishment, victims are more satisfied. It makes sense for those who transgress to feel, or at least display, remorse as it aids in their return to their communities and acceptance back into the moral majority.

But why should society care about the feelings of a person who has done harm, who has shown a lack of concern? In theory, the unrepentant transgressor's feelings shouldn't stop the rest of us from coming together to celebrate the reconfirmation of consensus around a shared value. And yet, it does make that celebration seem less satisfying, the whole experience less just. This is in part because the absence of remorse is yet another violation of social values. We can strip a criminal of every shred of agency, lock him in a hole, and even take his life, but if he is perceived to go to the grave never feeling even a fraction of the pain of his victims and their families, it just doesn't feel fair.* We want those who transgress to *feel* the pain they have caused.

Feeling like the person who has caused pain is now suffering isn't the only reason remorse feels more satisfying. Social scientists also suggest that we like remorse because of the hope of reform that

* See: Bill Cosby, Harvey Weinstein, Charles Manson … shall we go on?

piggybacks along with feeling bad. Emotions motivate action, and generally if people feel bad about something, they're going to take steps to *not* feel bad – they'll stop causing harm or promise to never do it again because they now see why it was so awful in the first place. These are the people we believe are safe to return to the community. At the same time, genuinely felt and/or perceived remorse also works to validate the victim as a human worthy of dignity and respect.

Self-punishment is one way of communicating remorse, leading observers and victims to believe that justice has been served and increasing the chances of reconciliation and return to harmony. Acts of self-punishment are deeply embedded in cultural, social and religious structures. These behaviours – observed in both religious and non-religious settings – can be anything from self-imposed poverty and fasting, to ritual flagellation and beatings, to weeping on national television and asking for forgiveness.

Evidence suggests that we are all primed to use pain to offset our own guilt and to restore a sense of fairness or justice. Pain, for example, is written into Christian theology – Jesus paid for the sins of the wicked world with his flayed flesh. Over the centuries since, the Roman Catholic Church developed multiple forms of 'mortification' of the flesh, such as putting a pebble in a shoe (called a 'scruple') or wearing a 'cilice' (described in Psalm 35:13 as a shirt lined with coarse hair) or even self-flagellation with whips. Their inspiration came from passages in the New Testament, such as when Paul writes in Romans 8:13: 'For if ye live after the flesh, ye shall die: but if ye through the Spirit do mortify the deeds of the body, ye shall live.' Even now, the Catholic faith finds that physical pain has a place. Every seven years, the hilltop Italian village of Guardia Sanframondi, all medieval stone, terracotta roof tiles and sunlit piazzas, plays host to the largest self-punishment rite in the Western world. These penitential rites stretch across a week-long festival featuring several processions celebrating the discovery of a statue of the Virgin Mary. The last day of the processions is referred to as the 'Day of Blood', where pain is the centrepiece of religious performance. Scores of men in white pointed hoods (a Catholic symbol way before the Klan claimed it), blood dripping from the lashes on their backs or bruises blossoming on their bare chest where they've ritualistically beaten themselves, use pain as purification of their sins.

We may not run for the whip every time we've crossed a line, but our inner voice may start nudging us in the direction of rebalancing by foregoing pleasures, such as skipping dessert, putting an embargo on spending, isolating yourself from friends. For more serious transgressions we may seek out some hurt, intentionally or not. This might show up as emotional self-punishment in the form of never-ending self-critical ruminations or intentional self-injury. Psychologist Brock Bastian found in an experiment that when study participants were reminded of an 'immoral deed', they were more motivated to experience and engage in physically painful activities. Further studies show that elective pain, outside of a religious context and in a laboratory, can indeed reduce feelings of guilt. Much of this appears to be happening without our conscious or cognitive effort. It makes sense: values are just that – highly *valuable* principles and beliefs that are central to our identity and self-concept. When we violate our values, we call into question our own moral compass. Are we a good person? If we are, why did we do something that goes against our principles? We have two basic options for solving this cognitive dissonance and reforming a stable self-concept. We can reconfirm the importance of the transgressed value by trying to make it right. That might mean opting in for the various forms of self-punishment, and/or repairing any damage. Our other option is to abandon the transgressed value, removing it from the self-formulated equation that calculates what you feel when you look at yourself in the mirror.

But just as self-inflicted pain might help you feel right with God and yourself, a little self-inflicted suffering may also go a long way in reconciling with others. A series of studies conducted by Melissa de Vel-Palumbo and colleagues at Flinders University in Australia found that self-punishment can indeed go a long way in restoring social harmony. Victims are more forgiving of self-punishing transgressors, and third-party observers find them to be more remorseful. Well, when it's done right. In these cases, self-punishment and expressions of remorse were accepted by the group as an implied endorsement of and recommitment to shared values and denunciation of their hurtful actions. Less trusting people, or those whose patience has previously been tested, might accept and believe that the perpetrator understands and genuinely regrets their hurtful actions, but be more hesitant in welcoming the transgressor back into the community, instead adopting more of a 'wait and see' approach. And those for whom trust is a sacred element, hard won and quickly lost, a show of remorse is just that: a show, meaningless without real, observed change.

The nature of self-punishment also impacts how it is received. In some cases more 'costly' or painful self-punishment is rewarded with a greater degree of forgiveness and welcoming back into the social fold. However, if it is deemed excessive, people might seriously question if the transgressor really does share the same understanding of values. Overcorrection just shy of excessive can also leave observers and victims feeling even more burdened. While the measure of what is 'too much' is culturally defined, in the Western world this might look like someone cutting themselves to atone for cheating on a loved one, or someone excessively apologising for minor infractions, expressing that they blame themselves, and self-criticising. In these cases, the wronged party now has to confront some internal dissonance: do they stay true to their value of compassion and reach out to now help them, letting go of hope for justice, or do they abandon their value of compassion in pursuit of justice? This type of overcorrection might cause friends or family to think the person is either disingenuous or has become too self-focused, so consumed in their own punishment that they have lost sight of what is important to the victim. It could also raise questions among parties about whether there is something else they feel an even greater guilt and need to atone, or that there is another motive – manipulation. This has the effect of denying the true victims justice, while simultaneously gaining sympathy.

A transgressor may also attempt to reconcile by self-punishment as a means of avoiding other, more costly means – such as actually repairing the harm they have done or facing their victim and apologising. For example, in a paper titled 'I've paid my dues', social scientist Lisanne van Bunderen from the Rotterdam School of Management at Erasmus University found that guilt offset by pain results in some individuals feeling less obligated to apologise or account for their transgressions. In fact, when given the option of how to atone, evidence suggests we take the easiest route, at least when no one is watching. Researchers investigated internet posts to a Yahoo! questions forum by users seeking advice on how to manage something they had done wrong. In these question-and-answer threads, the original poster could choose which answer they like the best while the community could also rate which answers they think are the best. The analysis revealed that posters tended to choose answers that were judged to be the most guilt-relieving, not necessarily the one that, as measured by the community ratings, was the most restorative.

We need empathy, but empathy can fail

The reason that pain can act as social glue – and social contagion – is empathy. Empathy is the mechanism that underpins emotional and physical connection. It is, as we saw in the previous chapter, based on our innate predictive modelling, the mental simulations that are happening every second we're alive. This enables us to not only actively meet our own needs, but also to feel what someone else might be feeling; neuroimaging studies consistently demonstrate that watching another person experience something activates similar neural activity in your brain. This ability is heightened when that person is a loved one or a family member: empathising with a stranger activates regions of the brain involved with mentalising, but empathising with a loved one activates regions involved in self-perception.

But the relationship between empathy and pain is a complex one. For example, can people who have never experienced pain empathise with someone in pain?

Steven Pete, the man who has never felt pain, admits that empathy has been 'a bit difficult for him'. Pete's wife suffers from 'suicide headaches', also known as cluster headaches – debilitating, crushingly painful headaches that occur every day for a period of time. Recently, his youngest daughter was diagnosed with the same thing. 'It's tough not feeling pain, but having your child who you love is in pain,' he said. 'And if you try to say anything, she looks at me like, "How do you know what I'm going through?" And I don't.' But Pete does know that it feels bad and that's probably enough – mirroring isn't the only mechanism behind empathy. In fact, a 2006 fMRI-based study compared the brain activation patterns of people with congenital insensitivity to pain (CIP) to controls during a series of photos of people in pain. Though the CIP patients demonstrated decreased activation in the visual regions, which researchers interpreted as reduced emotional arousal to viewing others' pain, they had increased activation in the areas of the brain associated with cognitively inferring others' emotional states. This implied that, even if they couldn't emotionally access another's pain by recalling instances of their own, that didn't stop them from empathising with it.

Even people with fully functioning pain systems can have a hard time conceiving the pain of others. Which is why we rely on cognitive tools, prediction and imagination to pick up the slack. Empathy – or something quite like it – is a skill that we can learn and practise. And we need to because, as much as we've talked about pain as a positive force, we have to recognise when it is not.

Pain can be isolating, depriving humans of contact with others, stripping them of their right to be heard, respected. Pain can make us vulnerable, terrifyingly so. Pain, especially chronic pain or violent, inflicted pain, can be socially destabilising and its experience can threaten a very human need to perceive the world as just, or at least bending towards justice. 'Across the board, when pain threatens these fundamental social needs, it has been shown to make the experience of pain worse … we need to understand, investigate and acknowledge that pain can *also* have dire costs in the social sphere, and that these costs deserve attention, both in research as well as in the clinic,' wrote Kai Koros, a psychology doctoral student, on Body In Mind, a University of South Australia website devoted to promoting a better understanding of clinical pain. 'Pain *is* social, thus managing the social context of pain *is* managing pain.'

Empathy is what cuts through that isolation. Some of this is through the shared experience of pain, especially for individuals suffering from complex, ongoing, frustrating pain conditions. Steven Pete said his adolescence was made infinitely better by the work his mother did in trying to find other families with children like him – not only did she actually find other families touched by this incredibly rare condition, but she did it before the internet too. Alphonse Daudet, a 19th-century French author whose writing about his life with the devastating effects of neurosyphilis was collected in a book called *In The Land of Pain*, described an evening spent with his fellow 'ataxics': 'Astonishment and joy at finding others who suffer as you do.' Many other people have found their pain tribe in social media, or even just felt less alone knowing that there's someone else on Instagram with a colostomy bag. These experiences bolster and create identity, reinforce an individual's worth to themselves and the world.

We also know, of course, that how medical caregivers treat people in pain impacts how they experience it. Empathy helps people feel better. Studies consistently demonstrate that when people feel empathy from their doctors, their pain decreases; when they feel heard by and connected to their doctors, their pain decreases.

And yet empathy alone is not enough. We can't depend on empathy alone to motivate us to take care of people in pain, on an individual level and societal one; we can't depend on our own stores of empathy to help people in pain because those stores are not limitless and, empathy might not be all that it's cracked up to be. '[O]n the whole,' Yale psychologist Paul Bloom suggests, 'it's a poor moral guide.'

In his 2017 book, *Against Empathy*, Bloom argues that empathy can lead people to make short-sighted decisions, or at least decisions that only benefit those they empathise with. In an example Bloom discusses in his book, researchers in a 1995 study told subjects about a 10-year-old girl – they called her 'Sheri' – who was suffering from a terminal disease. Sheri was on a waiting list for a treatment that could help her, but she was far down the list and there were many other children in front of her. But you can move Sheri up the list, participants were told, and save her life. Those who were only told that they could move Sheri tended not to, and said it wouldn't be fair to the other children on the list. But those who were first invited to think about how Sheri might feel – to empathise with her – moved her up. Empathy, here, steered them to make an emotional decision that would have benefitted one child and potentially harmed other children.

Empathy also doesn't always motivate compassion. In his own experiments, Bloom found that more empathetic people tend to push for harsher punishments for people they believe responsible for suffering. Other studies also demonstrate that it's empathy, not trait aggression or even perception of threat to self, that 'uniquely predicts aggression' in experiments involving the perceived distress of another. This held true even without provocation and seemed to only hinge on the other's distress not their actual danger. Bloom also notes that quite often the quickest way to incite hatred against an out-group is by weaponising empathy – demanding that people empathise with an innocent victim and punish not only the individual transgressor but also everyone like the transgressor. '[E]mpathy tilts the scale too much in favor of violent action,' he writes. Causing pain often, as we've demonstrated, has a veneer of morality.

Empathy is also not a boundless resource. In a 2012 study, researchers from the Université Paris Descartes noted that most studies of empathy in a medical setting focus on patient assessment rather than examining the cost of empathy to the caregiver. '[R]esearch suggests that being empathic has a psychological cost for health care professionals that can lead to "compassion fatigue". Moreover, reduced empathy may sometimes be necessary for physicians to fulfil their duties more adequately,' they wrote. A 2018 study in the *Journal of Compassionate Health Care* into the prevalence and levels of compassion fatigue in acute medical care nurses found that nearly half of the nurses were dealing with moderate to very severe levels. What's more, 37.5 per cent of the nurses assessed met all three criteria for a diagnosis of PTSD. Other studies have confirmed similar findings.

This isn't to suggest we should reject empathy full stop – that would be ridiculous, dangerous and downright stupid after all we've talked about so far. But it is to suggest both that we can do empathy better and that empathy cannot be the only way to answer pain more effectively. We can get better at practising empathy (for example, a feasibility study of medical students given empathetic validation training found that the training improved communication as well as both patient and student satisfaction), but when empathy fails – and it will – we need structural supports to keep the tent up.

When our humanity expresses its petty and protective nature, we need unbiased, professional measures to keep it in check. We need a class of social and medical care professionals who are educated about pain, who are versed in its mechanisms and trained to listen and who can pick up the slack. And we need empathy's cousin, compassion. As Bloom and others have argued, this is a much more useful, rational guide. 'In contrast to empathy, compassion does not mean sharing the suffering of the other: rather, it is characterized by feelings of warmth, concern and care for the other, as well as a strong motivation to improve the other's wellbeing. Compassion is feeling *for* and not feeling *with* the other,' wrote Tania Singer of the Max Planck Institute for Human Cognitive and Brain Sciences and Olga Klimecki of the Swiss Center for Affective Sciences. Compassion training, they say, works and it works better than simply encouraging empathy.

As much as pain can bring people together in moments of sharing and empathy, pain can also be a wedge between them. Daudet also wrote that he hated feeling like a burden to his family and invisible to the rest of the world. One of the most heartbreaking lines from Daudet's journal – and there are so many – comes near the end. He wrote that he didn't want to read weariness and boredom in the eyes of the people he loved. 'The only real way to be ill is to be by yourself,' he concluded.

It's true that no one will ever be able to feel our pain exactly, and that can feel isolating. It's true that pain can be destabilising, disruptive. But it's equally true that pain is a fact of the human condition that can bind us together, even when it's awful and ongoing and terrifying. If we can find joy, fellowship, love in pain together, then we can also find a way to survive it together. If we can use pain as a means of bonding and generating trust, then we can also find in pain a way to share this load without doubling it. We are not alone and pain shouldn't make us think that we are or want to be.

CHAPTER NINE

How to Tell Better Stories about Pain

We started this book with the claim that we all have a pain story. We've told a number of them: our own experiences, stories about pain that is ignored because it doesn't fit definitions of 'real' pain, or pain that goes untreated because of economics or politics; pain that is a source of joy, a way to inculcate resilience and pride, and a bond between people; pain that is endured, pain that is celebrated, pain that feels like justice. But a story isn't just the facts, a straightforward record of this happened and then this happened and then this. How stories are told has meaning but, more importantly, the act of telling a story, especially to yourself, makes meaning. So this is the last story we're going to tell: pain is an essential part of life – but suffering doesn't have to be.

We are all going to have the 'most' painful experiences of our lives, if we haven't already. Trauma forces us to re-evaluate ourselves and the world as we navigate it in a changed body. We might feel like everything we knew is wrong, including that we live in a just world where good things happen to good people and there is a reason for everything. We will question our own value and ability. This is the kind of pain that can cause lasting suffering, an ongoing state of hopelessness that shakes us to the core of who we are.

So how can we get to a place that respects and preserves the legitimacy of our pain without it becoming the centre of our daily existence? How can we keep the story of our pain from becoming the story of our suffering? And make the story not just about *not* suffering, but about triumph, about growth? The trick is in the telling.

In August 1980, Willie Stewart was an 18-year-old high school kid working in construction over the summer. An undefeated state champion wrestler, American football player and rugby player, Stewart was due to start college in the autumn. This gig, working 15 storeys up on the roof of the infamous Watergate building in Washington,

DC, was hot work but good enough money. On 15 August, while he was working on a cooling tower, Stewart's arm was entangled in a rope. The rope became wrapped around a blade of a huge, industrial-sized fan. The fan was on.

'It ripped my arm off from the elbow, clean down to the wrist,' he told us. 'I thought I broke my bones, until I saw my hand in front of me.' As in, no longer attached to his body. He lay on the roof; some of his fellow workers ran over to pick up what was left of his arm. 'From the bicep and the forearm was gone, clean, you saw white bone sticking out – it was perfectly clean. It had a rope tied around it that cleaned it perfect. But the arm was ripped off,' Stewart recalled. 'The bicep was slinking down the bone and I pushed my finger up on the bicep, and pushed it up into the cavity, and I wasn't bleeding that badly after that.'

If that was all that happened to Stewart that day, losing his arm to a rope and a giant fan would be enough. But the ambulance coming to help Stewart got stuck in traffic. His brother, working on the same site, put Stewart in the back of his truck and made it a little way before also getting stuck in the same gridlock. Stewart got out and ran the mile to the hospital, his right arm cradling what was left of his left.

'I ran to the emergency room – and I hear this from a lot of people with traumatic injuries – but I was relatively calm,' Stewart told us. He was bundled into surgery, where his arm was amputated just above his elbow.

Panic and shock, an overload of endogenous opioids and a sympathetic nervous system pumping out waves of adrenaline got him to the hospital and kept him from fully comprehending the horror of what had happened. But that only got him as far as the hospital. 'The pain usually comes quite a bit after the trauma and then – I wasn't a wrestler any more, I felt I wasn't an athlete any more,' he said. 'When you talk about pain, that pain isn't physical. Which, you know, if you ask almost anybody, it's pain that you have to go through to be able to describe it. You sense when people are in such despair, that's the pain … this pain goes on for a long time. This is like heartbreak.'

What followed was a year of depression, grief and loss. As Stewart struggled to mourn the person he had been and to come to grips with his new reality, he became more and more isolated and withdrawn. He wasn't an athlete any more, he thought, how could he be? All his promise was gone; he spent a lot of time alone in his basement.

His mother, however, remembering the athlete Stewart had only lately been, made him agree to run a 5K. So he did. It wasn't great, it wasn't fun. But then he did another. Within two years, Stewart was back at rugby, playing for the Division 1 team Washington, DC Rugby Football Club – he started on the D team and worked his way up to being captain. Meanwhile, he was still running, while adding other sports seemingly at random. He won a silver medal at the 2002 Paralympic Games in Nordic skiing. He spent 20 days kayaking the Grand Canyon, making him the first one-armed person to kayak the Colorado River. He's escaped from Alcatraz 15 times, run the Hawaiian Ultra Running Team's Trail 100-Mile Endurance Run (aptly shortened to 'HURT 100') twice. He's competed six times in the XTERRA World Championships off-road triathlons; in his first race, his prosthetic arm fell off during the cycling leg and he had to turn around to retrieve it.

In 2017, at the age of 55, Stewart became the first disabled athlete to complete all five challenges of the Leadman Series, culminating in the notorious Leadville Trail 100 race. His adaptive coping mechanisms during those races are a lot like those the athletes in Chapter 5 use: 'Even if you quit, you're still in pain, so you might as well go finish so you can go lay down.' (Like Adharanand Finn, he doesn't rely on his two pre-teen daughters for inspiration: 'They don't care,' he said, laughing. 'It's just some fantasy that we have that our kids will be proud of us.' *Truth.*) But the other coping mechanism Stewart uses, his own kind of superpower, is the fact that he's survived a traumatic accident that took his arm – and his identity – and lived to tell the tale. His own tale, the story where he's grown.

Many people will experience some kind of identity-rocking trauma. According to a 2017 report from the World Health Organization's World Mental Health Survey of 24 countries, 70.4 per cent of respondents reported at least one lifetime trauma; the highest proportions of these were rape, sexual assault, being stalked and the unexpected death of a loved one. However, researchers also average that about 58 to 83 per cent of people report growth following adversity and, although the real numbers depend on how one defines 'growth' and when people are asked, that's a lot of people.

Clinical psychologists Richard Tedeschi and Lawrence Calhoun coined the term 'post-traumatic growth' (PTG) in the mid-1990s to describe the positive changes in their patients following highly stressful events. PTG is different from resilience, the ability to 'weather through' and 'bounce back' or return to normal following trauma; it describes the positive changes that take an individual beyond where they were pre-trauma. The PTG literature organises growth into three domains: understanding of the self, others, and the world. Growth in understanding of the self includes gaining a greater recognition of personal strength. Through severe injuries and traumatising pain, survivors may discover just how strong they are and that serves as a great source of pride, reinforcing self-efficacy. Another common area of growth involves a shift in worldview characterised by reprioritising values, or discovering one's 'true' values; frequently, individuals report that material possessions drop in importance as a greater appreciation for health, wellness, friends and family come to the forefront.

Throughout the PTG literature, many survivors admit that recognising the fragility of life, or how vulnerable the body can be to injury, was really hard at first and a source of anxiety. But, over time, the anxiety eased and was replaced with a desire to be in the 'here and now', to not take anything for granted, especially friends and family. Another common, yet perhaps surprising, finding among those who report post-traumatic growth is gaining greater insight into the fact that people make mistakes. This in turn fosters a deeper sense of forgiveness and humanitarianism. In fact, respondents expressed greater desire to 'pay it forward', which may make the world feel more benevolent, and works to increase a person's compassion for the self, along with feelings of self-worth and self-efficacy.

Experiencing deep, traumatic pain would not be most people's chosen path to personal growth, and nor should it be. But it's crucial that we remember that pain doesn't have to be a roadblock on that path. Post-traumatic growth is all down to how we make sense of the things that happen to us – in other words, the stories we tell ourselves.

In the last few decades, research into how we make meaning shows that we have control over this process. For example, in his book, *The Secrets of Happy Families*, writer Bruce Feiler explored how family myth-making informed positive relationships and well-being. And what he found was that it's not so much the facts of the story as how it is told and retold. Take, for example, the time Linda's then 18-month-old son fell into the duck pond on a cold day. Would the story become a cautionary tale? 'This is what happens when you don't listen to Mummy!' Or would it become a funny story? 'And then he declared the pond a "poo-poo" place – I think he cursed it!' Or would it become a story about how this really scary thing happened, but in the end, we were all OK? 'Oh, watching him tumble into the pond was heart-stopping, but we fished him out super quickly and by the time we made it home – dripping wet – even he was laughing!*

As it turns out, Feiler says, the most effective family narrative is one that acknowledges hardships but highlights triumph and positivity over and through adversity. This oscillating narrative acknowledges that there was trouble, there was trauma, but that in the end we survived. This framing can also be useful in processing personal trauma, transforming it into *your* story.

Because trauma is signified by a sense of loss of agency, restoring a belief in one's will to act, to manage tasks and to pursue and reach goals is a powerful means to post-traumatic growth. One of the best ways of reclaiming that control is by being the one who gets to decide what your story means. From the moment we're born, we are looking for ways to make sense of the world around us. It's natural to ask 'why me?' and to try to find meaning in trauma. However, looking for meaning implies there is some inherent truth waiting to be discovered or that there could be a purpose or reason behind the trauma. There may not be. Instead, replacing 'finding' and 'looking' with 'making' and 'building' offers an opportunity to turn this into an active process, to rightly re-establish yourself as the definer of your reality.

There are multiple points of view from which to start a narrative, and researchers have identified the most common approaches when it comes to PTG. One approach is through perceiving the self as better

* The pond is a cursed place now.

off now by diminishing where we were. For example, some people appear to perceive growth by diminishing previous accomplishments in work, quality of friendships, activities in the daily lives and personal talents of their pre-trauma self. This can also be problematic: rather than diminishing the status of the pre-trauma self, some might ignore or discount the work they've done to recover. In this storyline, the post-trauma self is compared to the pre-trauma self or imagined peers and found wanting, as in: 'Before the accident, I was getting bigger bonuses every quarter. I've been back at work for a year and haven't got one.' Without adjusting for the impact of trauma, even big achievements might feel like not enough, leaving people feeling unmotivated and hopeless; that's not super helpful.

Another common approach is to start the narrative from the point of real or imagined others who are worse off than us. These are both examples of downward comparisons and they are important for coping, especially in the short term. Being able to say 'at least that didn't happen to me', or 'at least I didn't end up like those unfortunate people' and, ultimately, 'at least I didn't die', is a kind of mental safety net that we often need in the immediate aftermath of trauma.

Downward comparisons may seem insensitive, but research shows this approach can actually lead to more prosocial behaviours. To perceive ourselves as 'better off', we have to first imagine what life is like for someone who is in worse shape – we have to empathise. Then in coming back and comparing ourselves, our own pain not only seems more tolerable but often promotes a greater sense of compassion for self and others, and gratitude for life. Though not specific to downward comparisons, overall studies show that a big part of PGT is reporting increases in altruism and gratitude. A 2018 study published in *Innovation in Aging* found incidence of lifetime adversity was positively correlated to prosocialness, donating to charitable causes and engaging in volunteer activities. It is a beautiful illustration of how love for others can lift us in our darkest times.

Of course, the last thing most people want to hear after a traumatic incident is any sentence that starts with 'At least...' ('At least you've still got the other leg!'), as most of us need some time to be *in* the pain, to hurt and feel the intensity of that hurt. Advice from psychologists, grief counsellors and parenting experts all underscore the same thing:

acknowledge the emotion and let it live before trying to work on it. This holds true when confronting someone else's emotional state as well as our own; a key part of the narrative is acknowledging the trauma and its devastation before trying to get past it. Practically speaking, this means strategic planning, identifying resources and making a logistical roadmap of what you're going to do next. With each step, you're actively reminding yourself that you are alive, that it is your choice to live and your voice that will be the loudest.

But we can't do it alone. Social support and community engagement is essential for post-traumatic growth, especially if your roadmap involves moving from varying degrees of dependence to independence. This was the case for Stewart, who was pushed off the couch and out of the basement by his mother. 'It just took someone believing that I could do something, and then that belief creates courage,' Stewart told *Runner's World* in 2017. 'The biggest thing is you have to overcome fear, especially when you're newly injured. The physical activity, the act of going for a jog or a walk, will build on itself.'

And this is why a critical ingredient following a trauma is a trusted network of friends, family and professionals. PGT literature also overwhelmingly shows that growth following trauma is connected to forming deeper and closer relationships with trusted family and friends, which in turn is related to gaining personal strength and resilience. Encouragingly, one meta-analysis found that burn survivors were surprised at the outpouring of support and compassion, and this went a long way in validating their feelings of self-worth. Finding people who love and accept us always feels good, but following trauma this is especially important.

In fact, one study found that those who started new relationships after trauma reported higher post-traumatic growth compared to those who did not. This doesn't mean new relationships cause post-traumatic growth, but rather they reinforce self-acceptance and offer new opportunities to define how that trauma is going to be framed. It's even better if those new relationships are mutually beneficial. For example, a group of researchers and clinicians from University of New England in Australia wanted to learn what motivated peer mentors of at-risk youth. Through conducting semi-structured interviews with the mentors, they discovered it was the opportunity to rewrite their narrative: 'The experience of mentoring afforded opportunities to rewrite individual personal journeys of trauma through mentoring their at-risk peers, thus constructing a more positive self-identity.'

So much of our self-identity comes from our social interactions. For Stewart, how other people saw him was both hugely motivational and terrifying. 'I truly did not believe I lost my arm and I didn't think that I could go running,' he said, laughing. 'It wasn't that I didn't want to go running, it was that I didn't want someone to see me and if I tripped and fell they'd say it was because he had one arm! I didn't like my look in the mirror.'

Or in the eyes of other people. Influential sociologist Charles Horton Cooley's theory, 'the looking-glass self', describes how we make sense of who we are through our interactions with others; how we imagine others see us is integrated into our own perception of self. This is why it hurts so badly to be rejected and why, as studies show, the loss of once believed to be good friends following trauma or through chronic pain is so devastating. In fact, post-traumatic growth is linked to reframing the loss of friends post-trauma as opportunities to clarify the true nature of relationships, discovering and deepening relationships with those who stuck around, and letting go of the 'fair-weather' friends.

For Stewart, losing his arm abruptly put him on the outside of a social circle – the rugby players, footballer players – he'd grown up in. 'I was on the outside for two years, and once I got back in the inside, I never wanted to be on the outside again. I was really willing to take risks,' he said. 'The pain of isolation was worse than the pain of racing. Loneliness is painful.'

Stewart's struggles underscore the difficulties of navigating a post-trauma identity. Reframing, finding the positives, making a roadmap, writing a new narrative for yourself is easy enough to say, but it's far harder to practise, especially in the wake of a tremendous blow. Stewart said the fear of being on the outside was enough to make him take risks on the rugby pitch that, though they paid off and catapulted him up the ranks of the team, were dangerous. 'I pushed myself, I became the captain of one of the best teams in the United States,' he said. 'The only way I did that was taking extraordinary risks with my health and well-being against people twice my size. My desire on the field was more than my competitors', that's the strength of someone with a disability. I could overcome things with my desire … I wanted to be stronger, faster, I'd work out harder. I committed way more than someone who takes their skills and ability for granted.'

But, he said, he was still kind of an 'asshole' (he's definitely not now). 'The truth is when I got hurt I felt horrible about myself, I felt like an irrelevant human being. And I had looked at people with disabilities that way … the only way I learned to stop doing it was to have my arm chopped off,' he told us. It took him years to learn that his worth wasn't predicated on how hard he hit on the pitch, how fast he skied, how far he could run. 'Once you learn to love yourself, you can love others. And that's a hard lesson.'

Stewart's story also underscores a bigger point: post-traumatic growth doesn't happen overnight, it didn't happen the moment he crossed the finish line on his first 5K, or even over years. Studies show that growth is connected to the length of time since the trauma, so we have to be patient with our own progress. Age also makes a difference, not in whether growth is possible, but rather the nature of that growth. For example, younger people are more likely to see opportunities and possibilities they never before considered, while older folks gain a deeper appreciation for others. But ultimately, everyone is capable of post-traumatic growth.

'I think we're all capable of all kinds of stuff once we put our minds to it,' said Stewart. 'The human body – if you take one minute off and walk for a minute, you might be back. It's not over until it's over, and you have to decide when it's over.' Stewart's stubborn refusal to let trauma – suffering – be his only story means that the story of himself is one of pain and ultimately triumph. Everyone has a pain story, and how that story turns out – the plot twists and turns – depends on what we make of it. Stewart is the hero of his own story.

'I want to be racing into my seventies,' he told us. 'I want to drop dead racing, that's my hope. I want to drop dead with a number on.'

What we learned

In the introduction, we told you that we wanted you to know who we are, to get an idea of how our pain stories shaped our perspectives before asking you to into some potentially scary territory with us. And now, as we come to the end of our journey together, we want to share how our pain stories continued to unfold, how writing this book changed our experiences of the big pains, the little pains, the pain inflicted on us by other people, and pain inflicted on us by ourselves (or prosecco). Neither of us has lost an arm to a giant fan; our pain stories are not nearly so dramatic, so don't get your hopes up. But our stories – just like your stories – have meaning. They demonstrated to us that we can practise what we're preaching, that pain is negotiable. That we have the power to rewrite, even mid-sentence, an experience that so many people take as an absolute. Pain has the power to floor us, to sink us. But we have the power to raise ourselves up.

Margee: I can be brave

As I said in the introduction, my pain story didn't end with the liberation of my nerve track and fusion of my C5 and C6, it just started a new chapter. But at the time, I very much was of the opinion that my adventure in nerve pain *was* over and I could merrily go along my way. Turns out, it was that way of thinking that essentially guaranteed that in a few years I'd be contorting into awkward yoga poses just to get some relief once again. But to understand how I would get there, I've got to take us back in time a bit.

The six months between injury and surgery were not my finest. Chronic pain takes a toll on not just the person suffering it but also on those around them. By the end of November, I was struggling, to say the least, but I couldn't admit it. Everything was difficult to manage; showering and getting dressed, working at my computer, working late nights collecting data in a haunted attraction, basically anything that involved two arms working in tandem. I couldn't even brush my cats. Not surprisingly, I was a bit of a bitch to everyone. During a late-night work meeting, sitting in my chair like a really bad contortionist, someone asked why I was being so obstinate. I don't remember what we were talking about or how I was being difficult, but I have no doubt that I was. I remember snapping, 'I AM IN PAIN. THAT IS WHY.'

As we covered in Chapter 4, verbalising, giving voice to our pain is powerful, and in that moment I felt as though my voice had cut whatever strings were holding me together. Embarrassed, I ran out to my car, where I bawled my eyes out. But even crying hurt, and not in the sometimes helpful reprioritising of attention way – I couldn't drop my face into my hands without searing waves of fire burning down my arm. Throwing my arm behind my car seat headrest, my tears dried up and my sadness transformed into the negative emotion that unfortunately always seems the most accessible and easiest to hide behind: anger. But, like fear, anger can exacerbate pain. And really who, other than myself, was there to be mad at?

And I was mad at myself. We've talked a lot about the consequences of the Cartesian divide and how important it is to understand mind and body as an integrated whole – not in a spiritual or metaphysical way but in a very real 'this is how your nervous system functions' way. Yet, this continues to be one of my biggest challenges. My default is to ignore what my body has to say, and mostly only pay attention when it doesn't do what I want it to do. I've perfected tuning out rather than tuning in. Part of this is due to the fact that, when I would 'tune in', all I heard – all I *felt* – was bad.

In adolescence, cognitive behavioural therapy helped me to challenge my negative thinking and silence some of my ruminations, but it also taught me that my thoughts and feelings weren't always accurate reflections of reality. Rather than understanding this as an opportunity to get more connected with my body and the world around it – one of the primary therapeutic goals in CBT – I instead took it to mean that my mental health, at least in part, would always be a measure of how well I could convince myself of the 'right' feelings, of what I was 'supposed to' feel. Combined with never knowing how much of my feelings, or lack of feelings, was related to whatever antidepressant I was on, at some point I gave up on the idea that my feelings had anything true or useful to offer. (This, obviously, did not work out well.)

Affect, how we feel, is woven into who we are, what we think and do, and what we want in any given moment. Denying that reality, perceiving an annoying sensation as an obstacle to overcome, just makes things worse, as I've learned. Pain finds a way. I took up yoga and meditation in graduate school, learned the skills of interoceptive awareness and it helped tremendously with my anxiety. And yet, as time and life marched on, I would often forget this truth, that I needed to work to stay connected to my body. I'd slip back into a disconnected default mode, only to be reminded that I should listen to my body at the same point at which I also felt hopeless to do anything about it.

Months of enduring the pain in my left arm had dragged me down into the dark waters of negative valence where I sank to the affective sea floor. There, the Cartesian model ruled.

I shouldn't have waited as long as I did to get surgery. I should have called after an electromyography, which measures the health of motor neurons, revealed I'd lost about 20 per cent of the strength in my left arm, or when my friends and co-workers pointed out what I was clearly trying to avoid: I was in pain, and I needed help. But I didn't, not until the day I went to put my 16oz Frank's Hot Sauce back in the fridge and I dropped it.

In my medical sociology classes, I teach students the complex challenges that loss of ability brings: loss of agency and autonomy, loss of confidence, the feeling of helplessness, dependence and intense vulnerability. I have taught my students, many of them future nurses, how hard it can be to ask for help even when the need is so great, that the terror that accompanies feeling helpless and vulnerable can be far greater. But when my muscles gave out without warning, and I looked at my white-tiled kitchen floor now splattered with hot sauce and broken glass, I *felt* all of it. The fear, the confusion, the vulnerability. I also felt embarrassed, ashamed, sad. Not even anger could mask the intensity of emotion. This type of loss of agency is deep and it is powerful. How could I not know that my fingers were no longer strong enough to grasp and hold the weight of a bottle of hot sauce I picked up almost every day? I had become a stranger to my body, a body I didn't want to listen to, even as it suffered. I didn't tell anyone about the incident for over a year.

I called and scheduled the surgery – the earliest available, which was a little over a month away – that day. The day of surgery, I don't remember being scared. In fact, I thought the procedure sounded like a pretty amazing scene for a haunted attraction: my surgeon would make a small horizontal incision to the right of my throat about halfway down my neck, clamp everything to the side, and then fuse the C5 and

C6 with a bone plug and secure it with a plate. Because this approach doesn't require cutting any muscle tissue, when I woke up I was able to turn my head (and have a badass scar to boot). The procedure went exactly as expected and when I awoke I could immediately tell my nerves had been freed. After a night in the hospital I was sent home, the tingling and ache practically gone.

Physically, I was good to go. But a few weeks later, after the drugs had worn off and my friends and family left, I started having trouble sleeping. For six months I had slept in the same position, the only position, that didn't cause immense pain. Talk about pain as a corrective: any toss or turn in my sleep was met with a lightning bolt down my arm, reminding me to stay in check. As we talked about in Chapter 5, movement–pain relationships form quickly and are tough to break. Now that I could sleep in any position I wanted to, I was too scared. Scared that I might slip into a position in my sleep that would bend and break the hardware keeping my bones together, scared that the pain would come back.

At my follow-up appointment with my primary care physician, I told her I wasn't sleeping and why. Aware of my difficulty in trusting what my body has to say, she asked if I trusted my neurosurgeon. I said I did. She then told me to call him and ask how, if at all, I could 'ruin' his surgery. So I did just that. He very adamantly confirmed that no, I could not undo his surgery; there were four screws, a plate and a bone plug in there holding things together. Every time my anxiety started to rise I reminded myself of this, I did some breathing and eventually I was sleeping fine. But I still wasn't listening to, or trusting, the messages that were coming from my body.

But I was OK – mostly pain-free – for a long time. I went on my merry way, cervical spine issues in my rear-view mirror. Then, in the late summer of 2018 I felt those familiar tangled webs of tingles and twitches showing up in my right arm. It was a strange experience, to feel such a recognisable pattern of pain, what felt like the same pain show up in a different place, as though my personified collection of screaming fibres had just up and moved.

Linda and I were well into writing this book and, predisposed to not trust myself, a big part of me thought it could all be nocebo. I was reading and writing about pain every day; of course my body was going to manifest some of the sensations. So I distracted myself, broke out the old TENS unit, but mostly ignored the sensations again. Until I couldn't. Then living in Philadelphia, I found a new doctor and got back into the MRI machine; the findings revealed bulging disks at my C4 and C7. This time, I had lots of feelings. Would this be my life? Every few years fusing a few more vertebrae until I turned into the Tin Man? Thankfully, the disks were nowhere near as deteriorated as my C5 had been, I wouldn't need surgery anytime soon if I put in some work, hard work, with a physical therapist. My physical therapy prior to surgery had focused – rightly so – on pain relief, mostly trying to make space for my squashed nerve. But there were no soothing TENS or hot towels this time around. This time a team of physical therapists *brought the pain*, and some much-needed truth: I could not keep putting my body second, and I could not be a stranger to it.

Even as I wrote about the importance of interoception, of being connected and tuned in to the messages from all my bits and bones, I had progressively retreated from my body. As I wrote of the consequences of bubble-wrapped children, I was treating my entire body like a Fabergé egg. As I wrote of the healing found in sensation, I moved my body less and less. But just as it was true with kids, my adapting around protection had achieved the exact opposite. I was weak, really weak, in all the spots that I needed to be strong.

On my first day, like a puppeteer, my PT manoeuvred and adjusted my body into proper alignment against the wall. Looking at myself in the mirror, I felt a powerful dissonance: my body was straight but I felt like I was leaning at least 15 degrees to the right. That was the first surprise. Now in the proper alignment with a few rubber balls placed strategically behind my head and lower back, my PT told me to turn my head to the left and right 15 times. 'Really? That's it?' I thought, but as I looked to the right and started to turn, a tightness rose up through my chest, my face flushed and, before I knew it, I was crying – as in tears were welling in my eyes and started to drip down my cheek before I recognised how profoundly scared and sad I felt. The feeling was similar to when I had dropped the hot sauce: I was again the last to know what my body, what I needed, that I was hurting.

The nine weeks of physical therapy were the most difficult weeks of training I have ever done. But unlike training in roller derby or horseback riding, these exercises did not make me feel good, they made me feel weak. I used to be able to unload 50 bales of hay without blinking an eye, and now I could not even lift my head from the floor for more than 10 seconds. Intense emotions came up often, and I was routinely fighting to put the brakes on the wheels driving me down the safe, familiar roads towards anger and instead focus on awareness, acceptance and even compassion towards whatever my body had to say. A lot of people think that meditation practices are easy-breezy: as sensations and feelings arise, simply observe them without defence or judgement and allow them to float away like bubbles into the atmosphere. It's a nice visual, but for me, it often feels like walking to the edge of the ocean, watching the towering waves hurtling towards shore, and instead of running, trusting that your feet are planted solidly in the ground, that the waves will not take you.

Writing this book while going through this experience has been a mixed blessing. I can honestly say I have embraced all the tools we lay out here: educating myself, learning new words for sensation and putting them to use, practising mindfulness and meditation, connecting with others, and pumping my body full of electricity. But there were times when nothing would work and I would feel like, as Fiona Apple famously declared during her acceptance speech for the 1997 MTV Best New Artist award, 'This world is bullshit.' But I was writing a book on pain, I knew this world wasn't bullshit. Writing the stories of those we've interviewed, I was constantly reminded that mind and body are one complex, beautiful, integrated system, reminded how much we can endure, and of our capacity to heal.

I am stronger than I was when I started this journey with pain. The waves at the edge of shore aren't quite so towering any more, but even when they are, my feet are more firmly planted. And yes, the tingling nerve pain returns sometimes – under stress, after a particularly poor posture day, when sadness rolls into town. The small, and sometimes big vibrations skip and waterfall down through my arm and out to my fingertips, leaving a dull ache in their wake. I accept that this – and my crumbling bones – will always be a part of my life, part of me. My body is not my enemy, I am not my enemy. So I listen to my muscles, my skin and the chorus of sensation arising from my interoceptive networks, and when I hear 'pain' – instead of going for the ibuprofen and retreating into my mind – I move, I work on getting stronger, I go outside and lie on the grass, arms spread wide, and gaze up at the clouds. I *feel* as much as I can until the tingles fade into a watercolour of sensation.

Linda: What's on the other side of the wall

Two days after the race, I couldn't walk. I got stuck in the driver's seat of the car and had to call my husband from the driveway to come and pull me out. I couldn't lift the milk jug. I wondered how long it would take to install a chairlift in my house, to get me up the stairs. Would it be worth it? Maybe I should just sleep downstairs from now on. I Googled 'so sore' and came up with 'Delayed Onset Muscle Soreness'. Thanks, Google.

My case of Delayed Onset Muscle Soreness, or DOMS, was the result of a sunny Saturday spent running a 10K obstacle course called the Wolf Run in the English countryside. This, I should point out, was then unusual for me. Though I was a regular swimmer, I wasn't a runner and I've not been able to do a pull-up since, well, ever. I still can't. But I'd taken my youngest son to a new friend's house for a playdate ahead of the school year starting. That day, standing in their kitchen, the boy's parents explained that they were doing this Wolf Run thing – they'd done one a few months earlier and said it was really good fun – and were one person short of a team. Would I be up for it? The deadline for entering was that day.

Sure! Wait, what is it? Too late – I was committed. I was added to the WhatsApp group. My husband was, it should be said, not happy, raising the very sensible point that I could get seriously injured. I'd already been warned off running, trampolines and picking up anything heavier than a bag of flour by my doctor owing to the deplorable state of my pelvic floor after two children, he said, so this could not possibly be a good idea. The most I'd run in the last decade was for a train – and I hadn't made it. This, he pointed out, was a 10K run over uncertain terrain; judging by the website, it was going to be muddy, rocky, and hilly, punctuated liberally by obstacles to climb over or squeeze under. Oh, and there's swimming, he noted. You should be fine with that. Everything else is going to be awful. My older son, then seven, watched the promotional video with increasing alarm. 'Mommy,' he said in a low voice. 'What were you *thinking*?'

What I was thinking was that this would be a great opportunity for some real-life field research. It was one thing to talk about how to manage pain and meet people who have figured out ways to push through pain; it was another to actually use those techniques. And though we believed that pain could facilitate social bonding, this was a chance to witness that process close up, to see how it worked.

In the days before the event, texts from the group started to come fast and furious. Just how muddy were we going to get? And how much actual running was involved? How many lakes were we swimming across? And the water was going to be *how* cold? What should I wear? Did we even have a team name? Should we wear tutus? The night before the race, I watched a movie on the couch, my phone on silent. When I looked at it again, I'd missed 99 texts.

The next morning, I had a big bowl of porridge and didn't bother showering. I literally girded my loins with the special running shorts I'd bought to support my devastated pelvic floor, compression tights, and then the ugliest pair of nylon running shorts I'd been able to find at the grocery store. I taped my knees with kinesiology tape, slid on my hot-pink compression socks designed to manage my plantar fasciitis, and tied the shoelaces of my trail-running shoes as tight as they'd go. ('What if you lose a shoe in the bog?' my husband asked. 'I won't lose a shoe. That won't happen.' 'It might.') I was ready.

I did not lose a shoe in the bog. I did not fall and my pelvic floor stayed put. I did not pee in my pants. I did, however, experience searing, continuous sciatic pain down my left leg, a fierce cramp in the same leg, a cramp in my shoulder, a turned ankle, and the water. Oh god, the water. As it turns out, the most challenging part of the whole thing was swinging from a rope into water that was roughly freezing (well, not really, but it felt like it was). At that temperature, your body is trying to do the very serious work of keeping you alive, while your brain is busy panicking. I'd thought I was prepared because I understood the mechanics of cold-water shock. But I forgot how to swim. I doggy-paddled next to two of our group, gasping out a 'Are you OK?' directed at sort of everyone, really, every few metres.

In each of the moments where pain surfaced, what helped me continue on and ultimately through was the presence of other people. They served as both an inspiration – if they were running, well then, I could too – and as a conduit of self-criticism (meaning I didn't want to look bad in front of people I barely knew). This is what Margee told me is a situation of 'competing valence', a fight between which emotion is more salient: wanting to stop the pain or wanting to not let the team down or look like a tit. In the case when pride wins, it could be that this is an evolutionary response, part of a pack of responses that help people maintain social cohesion. It could also be that I am an overly sensitive person who likes other people to think she's far sportier than she is. It could be both. Whatever it was, it meant that in the battle between pain and peer pressure, however imagined, peer pressure won.

It was pain that bonded me to the people I was running with, but it was the people I was running with who made me ride out the pain. We travelled as a pack through the forest. We met our fellow runners, helped them over obstacles and up mud-slick riverbanks. When one of my teammates froze at the top of an A-frame 20 feet in the air, I helped her place her feet, uncurl her hands from around the ropes. When I couldn't muster the strength to pull myself up a muddy bank using the rope, one of my teammates hauled me out by my arm. We held hands crossing the bogs, steadying each other as our legs sank in mud over our knees. We did this over and over, reflexively checking at each obstacle to make sure we had all made it. At the last obstacle, a tower of hay bales to climb, we all reached the other side and, arm-in-arm, sprinted for the finish.

It was an absolute blast. But this sort of experience is also hard to communicate, or at least difficult to explain why you'd do it to people who aren't similarly inclined. Especially because it hurt. 'It looks awful, it really does,' my husband told me. This was after he'd seen the photos of all of us grinning in waist-deep, ice-cold mud, grinning as we pelted over the green grass, grinning as we slid belly-first down the giant water slide, and grinning as we stood together in front of the Wolf Run banner at the very end of the race. And it was after he'd seen us all, sore but still grinning like idiots, celebrating at the Indian restaurant the night after the race. It was also after we all signed up for the next one in eight months.

It's hard to communicate why hurling yourself over A-frames and through muddy bogs with a group of near-strangers is a desirable thing to do. But it was, not least because it was fun – when else do you get to play like a kid, with the same kind of abandon and willingness to get hurt in the pursuit of physical fun? It was worth the pain, even more so knowing that it was probably this pain – the thing we'd talk about over Indian food and beers that night, and that we'd joke about when they saw me limp through the school run on Monday – that would make us friends. Not for life, maybe, but more than we would be had we, say, just gone for the Indian food and the pints. Brock Bastian was right.

However, the Wolf Run was only part of a suite of pain-related activities that I was undertaking in the name of research. That was a one-off, an exercise that reinforced both pain's usefulness as 'social glue' and the importance of social bonds in managing pain. What I also wanted was an experience with pain that I could figure out how to manage and manipulate over time. Though I already deal with a few chronic situations, including monthly migraines, lower back pain from a recurring herniated disc and IBS, these were long-standing conditions and frankly, I'm bored of them. No amount of gaming has shifted them in the last two decades and, while I haven't given up, I've already reached acceptance. These things just *are* and though I don't enjoy them, I'm also not bothered by them. I also am not into BDSM or hook suspension, don't need another tattoo or piercing, and I've already got two children (see pelvic floor destruction). So that left sport.

It started with a thought that had occurred to me back in August 2016, on a camping holiday in the Lake District with my family. One day, we took the ferry up Lake Windermere to picturesque Waterhead, the kind of village filled with lots of Instagrammable moments and ducks (and lots of Instagrammable moments with ducks). It was raining and cold, and we were holed up at a pub with a view over the grey water. As we ate our chips, we watched swimmer after swimmer, some in wetsuits and some not, all red-faced and unsteady after hours of weightlessness, stumble up the short ramp from the lake and under an inflatable finish line. They looked dazed, as they should have

– they'd just swam 10 miles, from the bottom of the lake to the top. The water temperature was probably warmer than the air, but only just. I thought, 'I want to do that.'

But I didn't do that, at least not right away. Evidently, working your way up to a 10-mile swim is a lot harder than you'd think, especially when you fail to actually get into the water or do any sort of physical activity for more than a year after your decision. Luckily, I did know how to swim. I'd learned in my early twenties, taught by my then-boyfriend who had also been captain of our college swim team. He was a good coach and I'd loved the feeling of effortlessness that came with doing it well. But my will to swim – well, my will to get up at 5.30 a.m. to swim before work, plus the pool membership I had through his work – died with our relationship. So I stopped. For about a decade.

But this seed that was planted, watching the stumbling swimmers in Windermere, started to grow. Swimming couldn't be as stressful for my pelvic floor than other exercise, I thought. And as we began doing more work on the book, I started to think about what would happen if I did try to push myself. Could I reframe pain and fatigue as something else? In January 2018, I got into the pool. First, laps at the local leisure centre, and then, from May, at a coached class for adults looking to improve. The first few nights after the class, after timed sprints or continuous swimming at a pace, I'd climb in bed with arms like lead. A friend also swam at a lake nearby, so as the weather warmed up, I started getting up at 5.30 a.m. to go with her. And I started getting faster.

I found that most of the time, the first several hundred metres felt great – I was energised, relaxed, able to speed along at a clip that felt strong but possible. And then I'd hit a wall. Pain in my arms and legs, a deep ache and fatigue. The feeling that I just didn't want to do this any more, and really why should I?

Where the wall was varied according to where I was swimming. In the pool, the wall was after about 250 metres. In the lake, it was wherever my first lap ended, between 650 metres and 1,000 metres. Clearly, I thought, the wall was variable. What was also variable, I found, was the amount of time it took to get over the wall. Some days, it would be a few hundred metres of wrestling with myself. Sometimes it was even more, and I'd find myself wondering if I was actually swimming through mud instead of water. But what was really surprising was that on the other side of the wall was a place where the pain seemed to quiet down. It was like fatigue on mute. The uncomfortable feelings just became part of the fabric of this experience in a way that didn't bother me. The wall wasn't dependent on real physical information; the wall was me.

On 16 September 2018, I swam 4 kilometres in my favourite lake. On the third kilometre, both of my legs cramped at the same time, an awful hardness that started in my calf and spread to my whole hamstring. I turned on my back and floated until it passed. I swam my first race, in the artificial lake at Hever Castle, the family home of the ill-fated Anne Boleyn, on 23 September – 2.5 kilometres in remarkably cold water. I remember putting my face in the water at the starting horn and feeling my throat close with terror and panic.* My heart started beating wildly and I couldn't move my arms in sync, couldn't make my head turn to breathe. It was as close to a panic attack as I believe I've ever come, triggered by cold-water shock, the darkness of the muddy water beneath me, the churning bodies around me. I kept lifting my head out of the water, trying to see my way clear. But I kept moving – I knew that if I just kept trying to move forward, I could eventually calm down. And I did. By the time I rounded the first turn, my heart rate had evened, my breathing was regular. I came 24th out of a field of 83, eighth among the women, with a time of 49:31.

The next weekend was the Wolf Run, which proved that fitness as a swimmer does not easily translate to fitness as a runner. After we finished the race, drank the pints and signed up for the next one, I thought to myself, 'Next time, I'm going to be ready.' I already had my magic pants, the ones that kept my pelvic floor in place, so I started running on the days that I couldn't swim. Running was like swimming, except worse – where swimming compensated for the pain and fatigue with weightlessness, running couldn't. Swimming

* I've since learned: if you can, put your face in the water before the starting horn to get used to it – it's the act of cold water on your face that prompts the mammalian dive reflex, which in turns causes the feeling of not being able to breathe.

rarely made me sore any more; running made me wish I didn't have legs. But I started to go further, faster. I joined the local athletic club for weekly track sessions. These were and are gruelling; I am hopelessly outclassed, always chasing people faster than me – members of the team are nationally ranked,* have run the London Marathon in two hours and 35 minutes, completed the UTMB in under 30 hours.

Meanwhile, I started swimming longer distances. In April 2019, I swam my first 5K – that's 200 laps in the 25-metre pool – in a time of one hour and 35 minutes. My next 5K swim was in July, the London Dock2Dock, in one hour and 42 minutes. The water in the Thames was warm, brackish and tasted of petrol in places. On long swims, there's exhaustion, of course, but also chafing, headaches from wearing a swim cap for too long, bruises around the eyes from wearing goggles for too long. There's the soreness from the same movements over and over again, the deep ache in the muscles that starts to feel terminal. The cramps, both the perpetual threat of and the actuality. But I kept swimming, building up to what I was thinking of as my 'big swim' – the length of Coniston Water, a 5.25-mile lake next door to Windermere.

My super-sporty season culminated in three consecutive weekends at the end of the summer: I swam Coniston Water in three hours and 12 minutes, through rain, wind and sun; then a 5K charity swim in the Thames, in one hour and 34 minutes; and then ran a 10K road race in 50 minutes and 22 seconds. These numbers aren't amazing but they represent a significant amount of pain and fatigue – and, most importantly, pushing through them.

My new-found sportiness fundamentally changed my relationship not only with pain but also with my body. I listened more. Paying attention to things like when I should eat or drink, was I *really* tired or did I have more in the tank, could I push through the growing numbness in my toes, was entirely informed by the research we've done. And it worked. Naming exactly the muscle that hurt the

---

* I am also now nationally ranked: in 2019, I was 63rd in the UK for women in my age group (35 to 39), for the 5K, with a time of 23:12. This was entirely down to the fact that I was the *only* female in that age group to show up to that particular qualifying race.

most – the piriformis, thanks – meant I could localise the pain. Focusing on my breathing helped me control it. I also got better at understanding the difference between dangerous pain and discomfort pain; during the last push in writing this book, my right hip became unstable and could no longer power me up hills. I backed off running for a few weeks (admittedly, it took a call with my GP to convince me that I should); we are now in the negotiation phase – how much running is too much? If I promise to do Pilates later, will my hip just get me through this next kilometre? The pain of my swollen and blackened toenails was simply the cost of doing business; nobody died from a lost toenail.

But paying attention to the pain also meant I noticed when, for example, the pain of running eclipsed and even cancelled other pains, such as the pain of my regular migraines. The other thing we learned writing this book is that athletes do talk to themselves. And the kinds of things I say to myself wouldn't look out of place on a Pinterest quote board. It feels cheesy. It *is* cheesy. But it works: telling myself that I can do this makes me feel that I actually can. The act of reframing an experience gives us power over it. This skill is transferable, I found – helping me get through marathon nights of writing and rewriting a book.

For me, these weren't lessons I was ready for until after having two children. Childbirth pain was awful (no orgasms for me; you definitely would have read about that it in the introduction), but it eventually ended and I lived. The pain of running or swimming is even less intense, it ends more quickly, and I'll live. Plus, like Stewart said, even if you quit the race, you're going to be in pain. So you might as well keep running. If you're going to feel awful, you might as well feel awful *and* be running.

I have never quite had this kind of relationship with my body, ever. As a mother, my body hasn't totally belonged to me for the last nine years. When I was pregnant, I was literally sharing it. When I was breastfeeding, my body was a source of food for someone else. But even after that, even now, my body is not entirely mine. When my kids were smaller and wanted my attention, they would grab my face and turn it. 'Look, Mommy, look.' As they got bigger, they thought of me more like a mobile climbing frame. Now, most of the time my body is engaged in meeting other people's needs – on the school run, making dinner, doing laundry, finding lost toys. When I run or when I swim, however, the only person making demands on my body is me.

The nature of those demands has also changed. In my teens and twenties, exercise was about how my body looked – whether it was thin enough, mainly – and not about what it could do. I never thought about my heart. I never considered my lungs and the work they do. Now, however, I am inhabiting my body in a way that is new to me. I like what I can do with my muscles, I like how I feel, and I like knowing that I can run or swim through the pain. If I can do that, then perhaps I can keep from being sucked down by other pain.

What I've learned, though all of this, is that the things that used to signal to me to stop were just suggestions. I can ignore those suggestions. I've heard this from countless friends who have also become sporty in this stretch of their lives: when we were younger, pain was a clear signal to stop now and cut your losses; now, however, pain is negotiable. The wall *is* me and, like Willie Stewart says, it's not over until I say it's over. I have also learned to be patient, in part because I know that pain doesn't last forever and I'm willing to wait it out.

We can tell better stories

We won't lie. For all our new-found pain management skills, it would still be fantastic not to experience monthly migraines or dead arms. It would be fantastic not to experience fatigue or muscle soreness or churning guts or the emotional toll of being yelled at by strangers on the internet because you suggested that something they believe could maybe be considered in a different light. Some pain sucks.

But pain doesn't have to mean suffering. Though pain is inevitable, that doesn't mean we can't have a say over how it is felt and what it means. This means telling ourselves a new story about the things that happen to us, in a variety of mediums. Processing pain and trauma cognitively is essential but, remember, our construction of an emotion is formed not only from expectations and experience but also sensory input. We have to bring our body with us in our rewriting.

So we are encouraging you to *play*. Think of this as a mandate for exciting adventures. Whether it's through sport, adventure, acrobatics, BDSM or body rituals, engaging with your body is healthy and healing. We are sensation-generating machines, we might as well put them to work for us and have some fun. And remember that if some pain happens, that's OK. Pain reminds us that we are alive, that our actions have meaning, that we exist in a world of connections and other people and sensations and emotions.

Bad things will happen to us that will cause us to hurt. But that doesn't mean we have to let that be the end of the story. We are encouraging you to use the things we learned, from interoceptive skills and reframing to novel treatments and empathy, to make this experience less bad and sometimes good. We're practising these as much as we can, not just for ourselves, but for the people around us as well, people whose pain is being ignored because we just don't understand it or can't see it. We evolved to do better together – just holding someone's hand reduces pain. We need each other.

Pain is part of the story of who we are. And we can learn to tell better stories, and to make ourselves the heroes. We hope that by now – on the last page of this story – you'll agree with us.

Select References

Chapter 1

Barrett, Lisa Feldman. 2017. *How Emotions Are Made*. Macmillan.
Gosselin, Romain-Daniel, Marc R. Suter, Ru-Rong Ji, and Isabelle Decosterd. 2010. 'Glial Cells and Chronic Pain.' *The Neuroscientist*. https://doi.org/10.1177/1073858409360822.
Hartley, Caroline, Fiona Moultrie, Deniz Gursul, Amy Hoskin, Eleri Adams, Richard Rogers, and Rebeccah Slater. 2016. 'Changing Balance of Spinal Cord Excitability and Nociceptive Brain Activity in Early Human Development.' *Current Biology: CB* 26 (15): 1998–2002.
Hoemann, Katie, Fei Xu, and Lisa Feldman Barrett. 2019. 'Emotion Words, Emotion Concepts, and Emotional Development in Children: A Constructionist Hypothesis.' *Developmental Psychology* 55 (9): 1830–49.
Hoogen, Nynke J. van den, Jacob Patijn, Dick Tibboel, Bert A. Joosten, Maria Fitzgerald, and Charlie H.T. Kwok. 2018. 'Repeated Touch and Needle-Prick Stimulation in the Neonatal Period Increases the Baseline Mechanical Sensitivity and Postinjury Hypersensitivity of Adult Spinal Sensory Neurons.' *Pain* 159 (6): 1166–75.
Kleckner, Ian R., Jiahe Zhang, Alexandra Touroutoglou, Lorena Chanes, Chenjie Xia, W. Kyle Simmons, Karen S. Quigley, Bradford C. Dickerson, and Lisa Feldman Barrett. 2017. 'Evidence for a Large-Scale Brain System Supporting Allostasis and Interoception in Humans.' *Nature Human Behaviour* 1 (April). https://doi.org/10.1038/s41562-017-0069.
Mole, Tom B., and Pieter Mackeith. 2018. 'Cold Forced Open-Water Swimming: A Natural Intervention to Improve Postoperative Pain and Mobilisation Outcomes?' *BMJ Case Reports* 2018 (February). https://doi.org/10.1136/bcr-2017-222236.
Raja, Srinivasa N., Daniel B. Carr, Milton Cohen, Nanna B. Finnerup, Herta Flor, Stephen Gibson, and Francis J. Keefe, et al. 2020. 'The Revised International Association for the Study of Pain Definition of Pain.' *Pain*. https://doi.org/10.1097/j.pain.0000000000001939.
Wager, Tor D., Lauren Y. Atlas, Martin A. Lindquist, Mathieu Roy, Choong-Wan Woo, and Ethan Kross. 2013. 'An fMRI-Based Neurologic Signature of Physical Pain.' *New England Journal of Medicine*. https://doi.org/10.1056/nejmoa1204471.

Chapter 2

Ashton-James, Claire E., and Michael K. Nicholas. 2016. 'Appearance of Trustworthiness: An Implicit Source of Bias in Judgments of Patients' Pain.' *Pain*.

Carrico, Jacqueline A., Katharine Mahoney, Kristen M. Raymond, Logan Mims, Peter C. Smith, Joseph T. Sakai, Susan K. Mikulich-Gilbertson, Christian J. Hopfer, and Karsten Bartels. 2018. 'The Association of Patient Satisfaction-Based Incentives with Primary Care Physician Opioid Prescribing.' *Journal of the American Board of Family Medicine: JABFM* 31 (6): 941–43.

Casey, Logan S., Sari L. Reisner, Mary G. Findling, Robert J. Blendon, John M. Benson, Justin M. Sayde, and Carolyn Miller. 2019. 'Discrimination in the United States: Experiences of Lesbian, Gay, Bisexual, Transgender, and Queer Americans.' *Health Services Research* 54 Suppl. 2 (December): 1454–66.

Chen, Esther H., Frances S. Shofer, Anthony J. Dean, Judd E. Hollander, William G. Baxt, Jennifer L. Robey, Keara L. Sease, and Angela M. Mills. 2008. 'Gender Disparity in Analgesic Treatment of Emergency Department Patients with Acute Abdominal Pain.' *Academic Emergency Medicine: Official Journal of the Society for Academic Emergency Medicine* 15 (5): 414–18.

Cohen, Milton, John Quintner, David Buchanan, Mandy Nielsen, and Lynette Guy. 2011. 'Stigmatization of Patients with Chronic Pain: The Extinction of Empathy.' *Pain Medicine*. https://doi.org/10.1111/j.1526-4637.2011.01264.x.

Gross, Jacob, and Debra B. Gordon. 2018. 'The Strengths and Weaknesses of Current US Policy to Address Pain.' *American Journal of Public Health*, November, e1–7.

Hoffmann, D.E., and A.J. Tarzian. 2001. 'The Girl Who Cried Pain: A Bias against Women in the Treatment of Pain.' *The Journal of Law, Medicine & Ethics: A Journal of the American Society of Law, Medicine & Ethics* 29 (1): 13–27.

Hoffman, Kelly M., Sophie Trawalter, Jordan R. Axt, and M. Norman Oliver. 2016. 'Racial Bias in Pain Assessment and Treatment Recommendations, and False Beliefs about Biological Differences between Blacks and Whites.' *Proceedings of the National Academy of Sciences of the United States of America* 113 (16): 4296–4301.

Melzack, R., and K.L. Casey. 2014. 'Melzack & Casey Determinants of Pain 1968 from Original', April. http://dx.doi.org/.

Mezei, Lina, Beth B. Murinson, and Johns Hopkins Pain Curriculum Development Team. 2011. 'Pain Education in North American Medical Schools.' *The Journal of Pain: Official Journal of the American Pain Society* 12 (12): 1199–1208.

Pain Neuroethics and Bioethics. 2018. Academic Press.

Wailoo, Keith. 2014. *Pain: A Political History*. JHU Press.

Zuger, Abigail. 2013. 'Hard Cases: The Traps of Treating Pain.' *The New York Times*, 13 May 2013.

Chapter 3

Black, Nicola, Emily Stockings, Gabrielle Campbell, Lucy T. Tran, Dino Zagic, Wayne D. Hall, Michael Farrell, and Louisa Degenhardt. 2019. 'Cannabinoids for the Treatment of Mental Disorders and Symptoms of Mental Disorders: A Systematic Review and Meta-Analysis.' *The Lancet. Psychiatry* 6 (12): 995–1010.

Brody, Jane E. 2019. 'Millions Take Gabapentin for Pain. But There's Scant Evidence It Works.' *The New York Times*, 20 May 2019. https://www.nytimes.com/2019/05/20/well/live/millions-take-gabapentin-for-pain-but-theres-scant-evidence-it-works.html.

De Quincey, Thomas. 1885. *Confessions of an English Opium-Eater*. John B. Alden.

Donohue, Julie. 2006. 'A History of Drug Advertising: The Evolving Roles of Consumers and Consumer Protection.' *The Milbank Quarterly* 84 (4): 659–99.

Emmanouil, Dimitris E., and Raymond M. Quock. 2007. 'Advances in Understanding the Actions of Nitrous Oxide.' *Anesthesia Progress* 54 (1): 9–18.

Gallup, Inc. 2002. 'Decades of Drug Use: Data From the '60s and '70s.' Gallup.com. Gallup. 7 February 2002. https://news.gallup.com/poll/6331/Decades-Drug-Use-Data-From-60s-70s.aspx.

Goodman, Christopher W., and Allan S. Brett. 2019. 'A Clinical Overview of Off-Label Use of Gabapentinoid Drugs.' *JAMA Internal Medicine* 179 (5): 695–701.

Kaufman, David W., Judith P. Kelly, Deena R. Battista, Mary Kathryn Malone, Rachel B. Weinstein, and Saul Shiffman. 2018. 'Exceeding the Daily Dosing Limit of Nonsteroidal Anti-Inflammatory Drugs among Ibuprofen Users.' *Pharmacoepidemiology and Drug Safety* 27 (3): 322–31.

Lelorain, Sophie, Anne Brédart, Sylvie Dolbeault, and Serge Sultan. 2012. 'A Systematic Review of the Associations between Empathy Measures and Patient Outcomes in Cancer Care.' *Psycho-Oncology* 21 (12): 1255–64.

Mole, Tom B., and Pieter Mackeith. 2018. 'Cold Forced Open-Water Swimming: A Natural Intervention to Improve Postoperative Pain and Mobilisation Outcomes?' *BMJ Case Reports* 2018 (February). https://doi.org/10.1136/bcr-2017-222236.

Pizzo, Philip A., and Noreen M. Clark. 2012. 'Alleviating Suffering 101 – Pain Relief in the United States.' *The New England Journal of Medicine* 366 (3): 197–99.

Van Zee, Art. 2009. 'The Promotion and Marketing of Oxycontin: Commercial Triumph, Public Health Tragedy.' *American Journal of Public Health* 99 (2): 221–27.

Whipple, B., and B.R. Komisaruk. 1985. 'Elevation of Pain Threshold by Vaginal Stimulation in Women.' *Pain* 21 (4): 357–67.

Williams, J.T. 1997. 'The Painless Synergism of Aspirin and Opium.' *Nature*.

Chapter 4

Bergeron, Nicolas, Catherine Bergeron, Luc Lapointe, Dean Kriellaars, Patrice Aubertin, Brandy Tanenbaum, and Richard Fleet. 2019. 'Don't Take down the Monkey Bars: Rapid Systematic Review of Playground-Related Injuries.' *Canadian Family Physician Medecin de Famille Canadien* 65 (3): e121–28.

Felitti, V.J., R.F. Anda, D. Nordenberg, D.F. Williamson, A.M. Spitz, V. Edwards, M.P. Koss, and J.S. Marks. 1998. 'Relationship of Childhood Abuse and Household Dysfunction to Many of the Leading Causes of Death in Adults. The Adverse Childhood Experiences (ACE) Study.' *American Journal of Preventive Medicine* 14 (4): 245–58.

Green, Jarrod. 2016. *I'm OK! Building Resilience through Physical Play*. Redleaf Press.
Greenberg, David M., Simon Baron-Cohen, Nora Rosenberg, Peter Fonagy, and Peter J. Rentfrow. 2018. 'Elevated Empathy in Adults Following Childhood Trauma.' *PloS One* 13 (10): e0203886.
Jaskelioff, Mariela, Florian L. Muller, Ji-Hye Paik, Emily Thomas, Shan Jiang, Andrew C. Adams, Ergun Sahin, et al. 2011. 'Telomerase Reactivation Reverses Tissue Degeneration in Aged Telomerase-Deficient Mice.' *Nature* 469 (7328): 102–6.
Lee, Ja Y., Kristen A. Lindquist, and Chang S. Nam. 2017. 'Emotional Granularity Effects on Event-Related Brain Potentials during Affective Picture Processing.' *Frontiers in Human Neuroscience* 11 (March): 133.
Luthar, S.S., D. Cicchetti, and B. Becker. 2000. 'The Construct of Resilience: A Critical Evaluation and Guidelines for Future Work.' *Child Development* 71 (3): 543–62.
MacMillan, Harriet L., and Christopher R. Mikton. 2017. 'Moving Research beyond the Spanking Debate.' *Child Abuse & Neglect* 71 (September): 5–8.
Mathur, Maya B., Elissa Epel, Shelley Kind, Manisha Desai, Christine G. Parks, Dale P. Sandler, and Nayer Khazeni. 2016. 'Perceived Stress and Telomere Length: A Systematic Review, Meta-Analysis, and Methodologic Considerations for Advancing the Field.' *Brain, Behavior, and Immunity* 54 (May): 158–69.
Quoidbach, Jordi, June Gruber, Moïra Mikolajczak, Alexsandr Kogan, Ilios Kotsou, and Michael I. Norton. 2014. 'Emodiversity and the Emotional Ecosystem.' *Journal of Experimental Psychology. General* 143 (6): 2057–66.
Skenazy, Lenore. 2008. 'Why I Let My 9-Year-Old Ride the Subway Alone.' *New York Sun* 1: 2008–9.
'The Science of Neglect.' n.d. Center on the Developing Child at Harvard University. Accessed 3 December 2019. https://developingchild.harvard.edu/resources/the-science-of-neglect-the-persistent-absence-of-responsive-care-disrupts-the-developing-brain/.
Van der Kolk, Bessel A. 2015. *The Body Keeps the Score: Brain, Mind, and Body in the Healing of Trauma*. Penguin.

Chapter 5

Ambron, Elisabetta, Alexander Miller, Katherine J. Kuchenbecker, Laurel J. Buxbaum, and H. Branch Coslett. 2018. 'Immersive Low-Cost Virtual Reality Treatment for Phantom Limb Pain: Evidence from Two Cases.' *Frontiers in Neurology* 9 (February): 67.
Caspar, Emilie A., Axel Cleeremans, and Patrick Haggard. 2018. 'Only Giving Orders? An Experimental Study of the Sense of Agency When Giving or Receiving Commands.' *PloS One* 13 (9): e0204027.
Crum, Alia J., Peter Salovey, and Shawn Achor. 2013. 'Rethinking Stress: The Role of Mindsets in Determining the Stress Response.' *Journal of Personality and Social Psychology* 104 (4): 716–33.

Della Gatta, Francesco, Francesca Garbarini, Guglielmo Puglisi, Antonella Leonetti, Annamaria Berti, and Paola Borroni. 2016. 'Decreased Motor Cortex Excitability Mirrors Own Hand Disembodiment during the Rubber Hand Illusion.' *eLife* 5 (October). https://doi.org/10.7554/eLife.14972.

Gibson, Jonathan. 2019. 'Mindfulness, Interoception, and the Body: A Contemporary Perspective.' *Frontiers in Psychology*. https://doi.org/10.3389/fpsyg.2019.02012.

Harvie, Daniel S., Markus Broecker, Ross T. Smith, Ann Meulders, Victoria J. Madden, and G. Lorimer Moseley. 2015. 'Bogus Visual Feedback Alters Onset of Movement-Evoked Pain in People with Neck Pain.' *Psychological Science* 26 (4): 385–92.

Honzel, Emily, Sarah Murthi, Barbara Brawn-Cinani, Giancarlo Colloca, Craig Kier, Amitabh Varshney, and Luana Colloca. 2019. 'Virtual Reality, Music, and Pain: Developing the Premise for an Interdisciplinary Approach to Pain Management.' *Pain* 160 (9): 1909–19.

Johnson, Douglas C., Nathaniel J. Thom, Elizabeth A. Stanley, Lori Haase, Alan N. Simmons, Pei-An B. Shih, Wesley K. Thompson, Eric G. Potterat, Thomas R. Minor, and Martin P. Paulus. 2014. 'Modifying Resilience Mechanisms in at-Risk Individuals: A Controlled Study of Mindfulness Training in Marines Preparing for Deployment.' *The American Journal of Psychiatry* 171 (8): 844–53.

Kress, Jeffrey L.T. Slater. 2007. 'A Naturalistic Investigation of Former Olympic Cyclists' Cognitive Strategies for Coping with Exertion Pain During Performance.' *Journal of Sport Behavior* 30 (4): 428.

Roebuck, Gregory S., Donna M. Urquhart, Laura Knox, Paul B. Fitzgerald, Flavia M. Cicuttini, Stuart Lee, and Bernadette M. Fitzgibbon. 2018. 'Psychological Factors Associated With Ultramarathon Runners' Supranormal Pain Tolerance: A Pilot Study.' *The Journal of Pain: Official Journal of the American Pain Society* 19 (12): 1406–15.

Tesarz, Jonas, Andreas Gerhardt, Kai Schommer, Rolf-Detlef Treede, and Wolfgang Eich. 2013. 'Alterations in Endogenous Pain Modulation in Endurance Athletes: An Experimental Study Using Quantitative Sensory Testing and the Cold-Pressor Task.' *Pain* 154 (7): 1022–29.

Chapter 6

Abanes, Jane, Cynthia Hiers, Bethany Rhoten, Mary S. Dietrich, and Sheila H. Ridner. 2019. 'Feasibility and Acceptability of a Brief Acupuncture Intervention for Service Members with Perceived Stress.' *Military Medicine*, June. https://doi.org/10.1093/milmed/usz132.

Colloca, Luana. 2019. 'The Placebo Effect in Pain Therapies.' *Annual Review of Pharmacology and Toxicology*. https://doi.org/10.1146/annurev-pharmtox-010818-021542.

Dietrich, Arne, and Michel Audiffren. 2011. 'The Reticular-Activating Hypofrontality (RAH) Model of Acute Exercise.' *Neuroscience and Biobehavioral Reviews* 35 (6): 1305–25.

Hooley, Jill M., and Joseph C. Franklin. 2018. 'Why Do People Hurt Themselves? A New Conceptual Model of Nonsuicidal Self-Injury.' *Clinical Psychological Science* 6 (3): 428–51.

Lee, In-Seon, Christian Wallraven, Jian Kong, Dong-Seon Chang, Hyejung Lee, Hi-Joon Park, and Younbyoung Chae. 2015. 'When Pain Is Not Only Pain: Inserting Needles into the Body Evokes Distinct Reward-Related Brain Responses in the Context of a Treatment.' *Physiology & Behavior*. https://doi.org/10.1016/j.physbeh.2014.12.030.

McKenzie, Katherine C., and James J. Gross. 2014. 'Nonsuicidal Self-Injury: An Emotion Regulation Perspective.' *Psychopathology* 47 (4): 207–19.

Musial, Frauke. 2019. 'Acupuncture for the Treatment of Pain – A Mega-Placebo?' *Frontiers in Neuroscience*. https://doi.org/10.3389/fnins.2019.01110.

Reichl, Corinna, Anne Heyer, Romuald Brunner, Peter Parzer, Julia Madeleine Völker, Franz Resch, and Michael Kaess. 2016. 'Hypothalamic-Pituitary-Adrenal Axis, Childhood Adversity and Adolescent Nonsuicidal Self-Injury.' *Psychoneuroendocrinology* 74 (December): 203–11.

Weinberg, Anna, and E. David Klonsky. 2012. 'The Effects of Self-Injury on Acute Negative Arousal: A Laboratory Simulation.' *Motivation and Emotion*. https://doi.org/10.1007/s11031-011-9233-x.

Yang, Xing-Yue, Guang-Xia Shi, Qian-Qian Li, Zhen-Hua Zhang, Qian Xu, and Cun-Zhi Liu. 2013. 'Characterization of Deqi Sensation and Acupuncture Effect.' *Evidence-Based Complementary and Alternative Medicine: eCAM* 2013 (June): 319734.

Chapter 7

Ambler, James K., Ellen M. Lee, Kathryn R. Klement, Tonio Loewald, Evelyn M. Comber, Sarah A. Hanson, Bert Cutler, Nadine Cutler, and Brad J. Sagarin. 2017. 'Consensual BDSM Facilitates Role-Specific Altered States of Consciousness: A Preliminary Study.' *Psychology of Consciousness: Theory, Research, and Practice*. https://doi.org/10.1037/cns0000097.

Bastian, Brock, Jolanda Jetten, and Matthew J. Hornsey. 2014. 'Gustatory Pleasure and Pain. The Offset of Acute Physical Pain Enhances Responsiveness to Taste.' *Appetite* 72 (January): 150–55.

Lee, Kyung Hwa, and Greg J. Siegle. 2012. 'Common and Distinct Brain Networks Underlying Explicit Emotional Evaluation: A Meta-Analytic Study.' *Social Cognitive and Affective Neuroscience* 7 (5): 521–34.

Postel, T. 2013. 'Childbirth Climax: The Revealing of Obstetrical Orgasm.' *Sexologies* 22 (4): e89–92.

Richters, Juliet, Richard O. de Visser, Chris E. Rissel, Andrew E. Grulich, and Anthony M.A. Smith. 2008. 'Demographic and Psychosocial Features of Participants in Bondage and Discipline, 'Sadomasochism' or Dominance and Submission (BDSM): Data from a National Survey.' *The Journal of Sexual Medicine* 5 (7): 1660–68.

Rozin, Paul, Lily Guillot, Katrina Fincher, Alexander Rozin, and Eli Tsukayama. 2013. 'Glad to Be Sad, and Other Examples of Benign Masochism.' *Judgment and Decision Making* 8 (4): 439–47.

Sagarin, Brad J., Bert Cutler, Nadine Cutler, Kimberly A. Lawler-Sagarin, and Leslie Matuszewich. 2009. 'Hormonal Changes and Couple Bonding in Consensual Sadomasochistic Activity.' *Archives of Sexual Behavior* 38 (2): 186–200.

Wismeijer, Andreas A.J., and Marcel A.L.M. van Assen. 2013. 'Psychological Characteristics of BDSM Practitioners.' *The Journal of Sexual Medicine* 10 (8): 1943–52.

Chapter 8

Bastian, Brock, Jolanda Jetten, and Laura J. Ferris. 2014. 'Pain as Social Glue: Shared Pain Increases Cooperation.' *Psychological Science* 25 (11): 2079–85.

Cheng, Yawei, Chenyi Chen, Ching-Po Lin, Kun-Hsien Chou, and Jean Decety. 2010. 'Love Hurts: An fMRI Study.' *NeuroImage* 51 (2): 923–29.

Danziger, Nicolas, Kenneth M. Prkachin, and Jean-Claude Willer. 2006. 'Is Pain the Price of Empathy? The Perception of Others' Pain in Patients with Congenital Insensitivity to Pain.' *Brain: A Journal of Neurology* 129 (Pt 9): 2494–2507.

Davis, Joshua. 2014. 'Faculty of 1000 Evaluation for SUBJECT OF EDITORIAL EXPRESSION OF CONCERN: Experimental Evidence of Massive-Scale Emotional Contagion through Social Networks.' *F1000 – Post-Publication Peer Review of the Biomedical Literature*. https://doi.org/10.34 10/f.718431311.793500193.

Durkheim, Emile, and Karen Elise Fields. 1995. *Elementary Forms Of The Religious Life: Newly Translated By Karen E. Fields*. Simon and Schuster.

Klement, Kathryn R., Ellen M. Lee, James K. Ambler, Sarah A. Hanson, Evelyn Comber, David Wietting, Michael F. Wagner, et al. 2017. 'Extreme Rituals in a BDSM Context: The Physiological and Psychological Effects of the "Dance of Souls."' *Culture, Health & Sexuality*. https://doi.org/10.1080/13691058.2016. 1234648.

Kramer, Adam D.I., Jamie E. Guillory, and Jeffrey T. Hancock. 2014. 'Experimental Evidence of Massive-Scale Emotional Contagion through Social Networks.' *Proceedings of the National Academy of Sciences of the United States of America* 111 (24): 8788–90.

Linton, Steven J., Ida K. Flink, Emma Nilsson, and Sara Edlund. 2017. 'Can Training in Empathetic Validation Improve Medical Students' Communication with Patients Suffering Pain? A Test of Concept.' *Pain Reports* 2 (3): e600.

Losin, Elizabeth A. Reynolds, Steven R. Anderson, and Tor D. Wager. 2017. 'Feelings of Clinician-Patient Similarity and Trust Influence Pain: Evidence From Simulated Clinical Interactions.' *The Journal of Pain: Official Journal of the American Pain Society* 18 (7): 787–99.

Munson, Jessica, Viviana Amati, Mark Collard, and Martha J. Macri. 2014. 'Classic Maya Bloodletting and the Cultural Evolution of Religious Rituals: Quantifying Patterns of Variation in Hieroglyphic Texts.' *PloS One* 9 (9): e107982.

Olivola, Christopher Y., and Eldar Shafir. 2013. 'The Martyrdom Effect: When Pain and Effort Increase Prosocial Contributions.' *Journal of Behavioral Decision Making* 26 (1): 91–105.

Raspe, Heiner, Angelika Hueppe, and Hannelore Neuhauser. 2008. 'Back Pain, a Communicable Disease?' *International Journal of Epidemiology* 37 (1): 69–74.

Upton, Katherine Valentine. 2018. 'An Investigation into Compassion Fatigue and Self-Compassion in Acute Medical Care Hospital Nurses: A Mixed Methods Study.' *Journal of Compassionate Health Care* 5 (1): 7.

Vel-Palumbo, Melissa de, Michael Wenzel, and Lydia Woodyatt. 2019. 'Self-Punishment Promotes Reconciliation with Third Parties by Addressing the Symbolic Implications of Wrongdoing.' *European Journal of Social Psychology*. https://doi.org/10.1002/ejsp.2571.

Xygalatas, Dimitris, Ivana Konvalinka, Joseph Bulbulia, and Andreas Roepstorff. 2011. 'Quantifying Collective Effervescence: Heart-Rate Dynamics at a Fire-Walking Ritual.' *Communicative & Integrative Biology* 4 (6): 735–38.

Zhu, Ruida, Xueyi Shen, Honghong Tang, Peixia Ye, Huagen Wang, Xiaoqin Mai, and Chao Liu. 2017. 'Self-Punishment Promotes Forgiveness in the Direct and Indirect Reciprocity Contexts.' *Psychological Reports* 120 (3): 408–22.

Chapter 9

Douglas, Lesley J., Debra Jackson, Cindy Woods, and Kim Usher. 2019. 'Rewriting Stories of Trauma through Peer-to-peer Mentoring for and by At-risk Young People.' *International Journal of Mental Health Nursing*. https://doi.org/10.1111/inm.12579.

Fraire, N., M. Gardner, F. Infurna, K. Grimm, and S. Luthar. 2018. 'LIFETIME ADVERSITY AND PROSOCIALNESS: EVIDENCE OF ADVERSITY-DERIVED PROSOCIAL ATTITUDE AND BEHAVIOR.' *Innovation in Aging*. https://doi.org/10.1093/geroni/igy023.2591.

Gangstad, Berit, Paul Norman, and Jane Barton. 2009. 'Cognitive Processing and Posttraumatic Growth after Stroke.' *Rehabilitation Psychology* 54 (1): 69–75.

Grace, Jenny J., Elaine L. Kinsella, Orla T. Muldoon, and Dónal G. Fortune. 2015. 'Post-Traumatic Growth Following Acquired Brain Injury: A Systematic Review and Meta-Analysis.' *Frontiers in Psychology* 6 (August): 1162.

Kessler, Ronald C., Sergio Aguilar-Gaxiola, Jordi Alonso, Corina Benjet, Evelyn J. Bromet, Graça Cardoso, Louisa Degenhardt, et al. 2017. 'Trauma and PTSD in the WHO World Mental Health Surveys.' *European Journal of Psychotraumatology* 8 (sup5): 1353383.

Martin, Lisa, Michelle Byrnes, Sarah McGarry, Suzanne Rea, and Fiona Wood. 2017. 'Posttraumatic Growth after Burn in Adults: An Integrative Literature Review.' *Burns: Journal of the International Society for Burn Injuries* 43 (3): 459–70.

Powell, Trevor, Rachael Gilson, and Christine Collin. 2012. 'TBI 13 Years on: Factors Associated with Post-Traumatic Growth.' *Disability and Rehabilitation*. https://doi.org/10.3109/09638288.2011.644384.

Sawyer, Alexandra, Susan Ayers, and Andy P. Field. 2010. 'Posttraumatic Growth and Adjustment among Individuals with Cancer or HIV/AIDS: A Meta-Analysis.' *Clinical Psychology Review*. https://doi.org/10.1016/j.cpr.2010.02.004.

Acknowledgements

From both of us

Working on this book has been an adventure, from the very first conversations we had about our teenaged attempts to pierce our belly buttons to trying to edit (together!) in the middle of a global pandemic. We are so very grateful to the people who shared their stories, perspectives, and research with us for this book – regardless of whether it made it into the book, your work and your voices were enlightening and motivating, and inspired us to reconsider how we understood pain. We are also thankful to the writers, researchers, and scientists who have explored the experience of pain before us; every word here is built on their hard work, curiosity, eloquence, and even their mistakes.

We'd like to thank our literary agent, Tisse Takagi (go #TeamPain!), and everyone at The Science Factory, and The 'A-team' at Bloomsbury – Anna MacDiarmid, Angelique Neumann, Alice Graham, Amy Greaves, and Jim Martin. We're so pleased that you all saw the need for a book like this and so hopeful that we've lived up to your expectations.

From Margee

The process of writing this book has indeed been an adventure, one that I would not have completed without the support of friends and family. My biggest thanks goes to Linda; I experienced the real magic of creative collaboration and the many moments where our ideas came together as we created something entirely new brought me so much joy. Writing together has taught me more about communication, compromise, and partnership than I could ever have imagined, and for that I am eternally thankful.

To my wife Amy, who doesn't turn away from my pain but asks to help shoulder it, who manages to bring light even into the darkest moments, and whose love is the best cure, thank you.

And to my parents, I am so incredibly grateful for your unconditional love. Your faith in me continues to lift me when I am low and support me when I fear I'll fall.

From Linda

Pain is a shared experience, and so too is the pain of writing a book! In this case, I got to share it with someone whose brilliant brain and unique perspective made putting this together a wonderful process of discovery. This book was Margee's brainchild. Thinking through the complex layers of what pain means together and seeing the science and stories with the help of her brilliant analogies and coming up with a way to explain and express them here has been a wonderfully collaborative experience. I'm grateful for it and for her!

I'd also like to thank my friends and family who shared their painful experiences with me as we wrote this; those stories didn't make it into the book, but did very much inform my process and broaden the way I think about pain.

A big thank you to my mom and stepdad, whose unwavering support is more than I could have asked for, and to my kids, whose arrivals into this world were immensely painful but whose existence in it continues to be wondrous. And, of course, I owe a huge debt of gratitude to my husband, Chris McRobbie, for, well, everything. Thank you.

Go #TeamPain!

Index

Notes are denoted by a letter n following the page number.